CW01628293

Sustainable Finance

"The damage banks and investment companies are doing to our world and our climate is nothing short of criminal. We urgently need to turn this around and use the power of money for good. Molly Scott Cato explains how with the clarity and insight she has gained from her unique background as both an economics professor and a politician regulating Europe's financial system."

—Caroline Lucas, *Green Party MP*

"For too long 'the economy' and 'the finance' flowing through it have proceeded as if their own self-defined laws are immutable and detached from people and the planet we inhabit. Molly Scott Cato has been at the forefront of the sustainable finance agenda and has written a must-read primer for anyone interested in understanding how to align the financial system with climate objectives. With authentic experience of the 'cut and thrust of political negotiations', Molly not only offers first-hand perspective on why progress on this vital agenda has been slow but offers compelling proposals for what to do about it."

—Frank van Lerven, *Senior Economist, New Economics Foundation*

Molly Scott Cato

Sustainable Finance

Using the Power of Money to Change the World

palgrave
macmillan

Molly Scott Cato
Green Economics, Business School
University of Roehampton
London, UK

ISBN 978-3-030-91577-3 ISBN 978-3-030-91578-0 (eBook)
https://doi.org/10.1007/978-3-030-91578-0

© The Author(s), under exclusive license to Springer Nature Switzerland AG 2022
This work is subject to copyright. All rights are solely and exclusively licensed by the Publisher, whether the whole or part of the material is concerned, specifically the rights of translation, reprinting, reuse of illustrations, recitation, broadcasting, reproduction on microfilms or in any other physical way, and transmission or information storage and retrieval, electronic adaptation, computer software, or by similar or dissimilar methodology now known or hereafter developed.
The use of general descriptive names, registered names, trademarks, service marks, etc. in this publication does not imply, even in the absence of a specific statement, that such names are exempt from the relevant protective laws and regulations and therefore free for general use.
The publisher, the authors and the editors are safe to assume that the advice and information in this book are believed to be true and accurate at the date of publication. Neither the publisher nor the authors or the editors give a warranty, expressed or implied, with respect to the material contained herein or for any errors or omissions that may have been made. The publisher remains neutral with regard to jurisdictional claims in published maps and institutional affiliations.

Cover illustration: © Melisa Hasan

This Palgrave Macmillan imprint is published by the registered company Springer Nature Switzerland AG
The registered company address is: Gewerbestrasse 11, 6330 Cham, Switzerland

To the millions of young people around the world who have risen up to demand that we take urgent action on the climate crisis

Acknowledgements

For their pioneering leadership on the sustainable finance agenda, I would like to thank Paddy Arber, Sandrine Dixon-Declève, Nick Robins, Nancy Saich, and Martin Spolc.

For their spiritual leadership during a time of crisis, I would like to thank His Holiness the Dalai Lama and His Holiness Pope Francis.

For their personal support for my work on sustainable finance and for their patience, I would like to thank Cory Fletcher, Francisco Padilla Olivares, and Andrew Bell.

And for political solidarity, I would like to thank Pervenche Beres, Sven Giegold, Sirpa Pietikäinen, and Paul Tang.

Prologue

I come at this topic from two distinct perspectives, having had two distinct careers. First, I approach the question as a Professor of Green Economics. This role is about understanding the very complex detail that has grown up around the sustainable agenda and, more widely, the green economy. I have done my best to understand what is a vast policy agenda and summarize it in an accessible way. I can only apologize for errors and omissions and ask for understanding because this agenda is moving very rapidly, as is the whole debate around the climate crisis, and indeed the crisis itself.

My second perspective is that of somebody who has been involved in the cut and thrust of political negotiations over sustainable finance as a Member of the European Parliament. This is, of course, where knowledge meets power, and in ways that sometimes left me weeping with frustration. It is horrifying to see politicians sacrifice the lives of future generations just to protect the profits of the car industry or, I could not help suspecting at times, for their own financial advantage. I hope that perspective is helpful in explaining why our action on the most dangerous crisis we have ever threatened as a species has been shockingly slow.

The sustainable finance agenda has been driven by what I have seen described as a 'coalition of the unlikely' and that is certainly how it felt to me when I found myself getting on so well with the executives of multinational insurance companies and being invited to speak at City of London events. It is an issue that separates the bold from the timid, the public-spirited from the self-interested, those with imagination from those who believe what they are told. I know that is a tendentious way of describing it but I also know which side of each of those divisions you would like to find yourself.

The focus of attention in discussions about sustainable finance is heavily on climate and I have followed that tendency in this book, inevitably because the breakdown of our climate is confronting us daily. And for the multiple other environmental crises waiting in the wings—from the loss of soils to the breakdown of the nitrogen cycle, from the looming extinction of pollinators to the choking masses of waste plastic—the financial regulations and policies designed to address the climate crises either already encompass them or can be used as templates for regulation to address them. The issue of climate is breaking the ground in introducing sustainability factors into financial regulation and the operation of financial markets as well as banking and wider economic policy: similar policy on other environmental—and indeed social—crises will follow.

Saving life on earth has never been a minority interest and most of us have grandchildren or other young people whose lives we value and seek to protect. If, like me, you have always been suspicious of how finance works and what financiers are up to, I encourage you to empower yourself by exploring all the ways money can be used as part of the struggle to save life on earth.

Stroud
September 2021

Molly Scott Cato

Contents

About the Author

Molly Scott Cato is Professor of Green Economics at Roehampton University. Between 2014 and 2020 she represented South West England and Gibraltar in the European Parliament where she was a member of the Economics and Monetary Policy Committee and was the Parliament's rapporteur on sustainable finance. She later helped negotiate the regulations on low-carbon benchmarks and mandatory disclosure.

Molly studied politics, philosophy, and economics at Oxford University and later gained a doctorate in economics from Aberystwyth. She joined the Green Party in 1988 and is the party's spokesperson on economics and finance.

Aside from her work as an economist. Molly's areas of special interest include land ownership and food production; renewable energy, especially when it is owned by local communities; cooperatives; and issues concerned with peace and opposing nuclear weapons and nuclear power.

Abbreviations

CCC	Committee on Climate Change
COP	Conference of the Parties
CPF	Carbon Price Floor
ECB	European Central Bank
EIB	European Investment Bank
ESG	Environmental, Social, and Governance
ETS	Emissions Trading System
G7	The 'Group of 7' high-income countries: Canada, France, Germany, Italy, Japan, the United Kingdom, and the United States
G20	The 'Group of 20' high-income and large and rapidly growing global economies: Argentina, Australia, Brazil, Canada, China, France, Germany, India, Indonesia, Italy, Japan, Mexico, the Republic of Korea, Russia, Saudi Arabia, South Africa, Turkey, the United Kingdom, the United States, and the European Union
GCF	Green Climate Fund
GEF	Global Environment Fund
GND	Green New Deal
GRI	Global Reporting Initiative
HLEG	EU High-Level Expert Group on Sustainable Finance
IEA	International Energy Agency
IMF	International Monetary Fund
MDB	Multilateral Development Bank
NFRD	Non-Financial Reporting Directive
NZC	Net Zero Carbon
OECD	Organisation for Economic Cooperation and Development
QE	Quantitative Easing

SDR	Standard Drawing Rights
SFDR	Sustainable Finance Disclosure Regulation
TCFD	Taskforce on Carbon-Related Financial Disclosure
UNEP	United Nations Environment Programme
UNFCCC	United Nations Framework Convention on Climate Change

List of Figures

Chapter 3

Chapter 4

Chapter 5

Chapter 6

List of Tables

CHAPTER 1

Why Sustainable Finance? Why Now?

Abstract After the failure of the UN climate talks in Copenhagen in 2009 a small group of finance experts decided to find ways to use the power of finance to tackle the climate emergency. Through developing ideas like 'stranded assets' and 'carbon bubble' they managed to convince financiers and bankers of the risks of not addressing the climate and environmental crises. This was the birth of the sustainable finance agenda. But it coincided with citizen efforts to channel finance away from destructive activities, especially through divestment from fossil fuels and avoiding banks that were investing in destructive activities.

Keywords Copenhagen climate talks · Stern Review · Stranded assets · Carbon bubble · Divestment

1 All Hands to the Pumps

It is now widely agreed by global governments and international organizations that climate change is the most urgent challenge facing humanity, an existential challenge not just for our way of life but for our very survival as a species. And while climate is the crisis that dominates our lives and this book, the breakdown of ecological systems is at least as threatening

© The Author(s), under exclusive license to Springer Nature Switzerland AG 2022
M. Scott Cato, *Sustainable Finance*,
https://doi.org/10.1007/978-3-030-91578-0_1

to our species in the long run—and in very specific instances in the very short run too.

I should say that I do not share the view that nothing happens without finance, or even that money is the most important aspect of solving the climate crisis. Indeed, it is clear that—whether we think about countries or individuals—those with more money at their disposal are contributing more to the problem of climate change than those with less money. A decision about whether to drive a 4 × 4, drive an electric car, or catch the bus is a decision driven by culture, the availability of alternatives, and the advertising industry as much as it is driven by one's access to disposable income.

But the global economy is organized along capitalist lines. Most of the goods and services that contribute negatively or positively to the climate crisis are bought and sold in private markets. The companies that sell them and the consumers who buy them are enmeshed in an economic system where these activities are facilitated by finance. And the costs of the goods and services are subject to policy decisions made by governments, who can create positive or negative incentives for both producers and their financiers to take decisions that exacerbate or help to ameliorate the climate crisis.

So when I write about sustainable finance I am not seeking to portray financiers as the white knights of the climate and ecological crises, riding in to solve the problems of humanity. In many ways my perspective is the reverse of this: in my view, the quest for profits above all else has significantly contributed to the destruction of our environment and my work as a Professor of Green Economics has always been intended to make that explicit and seek to counter it.

There are two reasons why I have worked much more closely with financiers in recent years than I ever expected to: both are pragmatic. First, we are in a climate *emergency* and we need to use all powers at our disposal. Whatever my personal views about the ethics and social injustice impacts of global finance, I have been unable to avoid the conclusion that this sector, as well as all others, must be enlisted in the struggle to keep the global climate within livable limits.

Secondly, in my role as an MEP, I was a member of the Finance and Monetary Policy Committee of the European Parliament. I was determined to use that power to curtail the power of the fossil fuel sector and encourage the necessary shift of global finance away from environmental destruction and towards the sustainability transition. As anybody engaged in front-line politics will tell you, it is not about black and white but mostly about negotiating the best you can in the grey areas. For me, sustainable finance is just such a grey area. But I did meet some people working for insurance companies, banks, and investment houses who had understood the vital role they could play and had informed themselves and acted with courage and intelligence in a way I found deeply impressive. Others, needless to say, continued to block progress and seem immune to the distress of those living in Bangladesh or Mozambique and the more distant voices of our grandchildren, their grandchildren. And of course I also had the pleasure to work with some wonderfully sharp and dedicated advisors for NGOs who have picked up and run with the sustainable finance agenda since Copenhagen.

This agenda is moving fast. Across the world, governments and bankers are collaborating to discuss how to use the regulatory tools at their disposal to turn the tanker around or, in the words of Herman Daly, to convert the aeroplane to a helicopter while it is still in flight. (a Greener version of his analogy would be welcome, of course). Of all the areas I have worked on, this is the one that has made the most positive change in the shortest time so part of the reason for writing this book is to share that as a note of optimism. But it is equally important that I ring alarm bells about greenwashing, backsliding, and hypocrisy on the sustainable finance agenda, as I will do in the following chapters.

2 From Copenhagen to Paris—By Train

I date the expansion of the sustainable finance agenda to the failure of the Copenhagen climate talks in 2009. Up to that point, well-meaning environmentalists—myself included—believed that we could use the power of science. We thought that once an international consensus on the risks to humanity from climate breakdown could be established, governments would mobilize to take urgent action. But the talks dissolved in chaos and disappointment with no clear way forward and no agreed targets for action.

Rather than sinking into despair, environmentalists began to ask where the power lay to make change. They had put their faith in science but that had not been enough. Climate talks had proved again that politicians tend to operate according to the adage 'I don't know much about science, but I know what I like'. Rather than science-driven policy they tend to prefer policy-driven science. So what was driving their refusal to shift towards urgent action when the facts about climate breakdown were so clearly proven? The conclusion was that it was corporations and their financiers who were blocking progress—driven by balance sheets and the bottom line.

So the environmentalists sought to reframe the way international business works so that it encompasses the environment, redefining the bottom line so that it includes environmental and social impacts and valuing assets in such as a way that the damage that will be caused to them by climate and ecological damage will be included in their valuation. The details of how they did this are covered in the following section. First, I need to acknowledge another important milestone: the Stern Review.

The Stern Review was led by Nicholas Stern, written by Treasury economists, commissioned by the UK's Labour Chancellor Gordon Brown and delivered in 2006. Its remit was to explore and elucidate the economic challenges posed by climate change and to provide recommendations. It was a landmark moment because—for all the limitations on this approach—it was the first time climate change had been framed by a national government as an economic problem.

Nicholas Stern's famous conclusion was that climate change is 'the greatest market failure of all time', an earth-shattering but also troubling conclusion. There is no question that a market system that undermines its ability to make products and may destroy not just the livelihoods of those who are needed to buy them but perhaps eliminate them altogether must be considered to be failing. But this immediately gives rise to the question: is the global market system essentially flawed and in need of fundamental redesign, or can we change the incentives driving market activity? Stern took the latter view and that is also the motivation for activity in the field of sustainable finance. He concluded that the primary source of market failure is the problem of public goods, where the market price of a good that has been produced does not include the social benefits or costs that arise from its production. Those who have caused climate breakdown are not required—within the market system—to pay for its consequences, so that problem must be dealt with politically. Since the use of the global atmosphere is not costed, this 'free good' is overused as a result of the emissions of CO_2 in production and transport—hence the focus on 'putting a price on carbon' we will explore in more detail later.

Perhaps the most important conclusion from the analysis was that climate action sooner is much cheaper than climate action later: delay in itself builds in additional and unavoidable costs. Given that this conclusion was reached a full 15 years ago it is a tragedy for us all that it was not heeded with more alacrity. The accelerating crisis is already bearing out this conclusion, as crises are interconnecting, intersecting, and reinforcing each other in ways that are causing increasing concern, for example the feedback loops now driving accelerating climate breakdown, such as the melting of polar ice-caps and the arctic tundra, leading to the release of massive amounts of CO_2 into the global atmosphere.

The Stern Review led to the UK government passing the Climate Change Act 2008, the first government in the world to pass such a law. It set binding targets for CO_2 reductions—leading to a Climate Change Levy—a prototype carbon tax (see more in Chapter 4) and a Committee on Climate Change (CCC), independent of government, to monitor progress to achieve these targets. Although its proposals for environmental taxes and massive public investment have not been followed up either in the UK or elsewhere with the urgency Stern proposed, this report set the scene for climate moving from being a concern in the environmental portfolio to one in the economic portfolio, and hence for the development of the sustainable finance agenda. (Chapter 4 includes more details of policy developments in various countries.)

3 Stranded Assets and the Carbon Bubble

Now we need to return to the story of the disappointed climate campaigners and how they reframed the narrative around the need for climate action after the failure of the Copenhagen COP. The Stern Review had made clear the massive costs of failing to take climate action but how could that theoretical understanding be conceptualized in a way that made it tangible for those responsible for the flows of vast amounts of finance through the global economy? How could it be made to feature in the bottom line of the world's corporations?

Two concepts were defined and mobilized to achieve the necessary shift in understanding of which the first and most important is the idea of 'stranded assets'. The idea of a stranded asset is not a new one and has been used by accountants to describe assets that, due to economic or social changes, no longer have the valuation that was previously placed on them. In the context of sustainable finance, a stranded asset is one which has lost its value either as a direct result of the climate crisis or due to policy or demand changes that means it cannot be exploited as was previously thought possible. The concept of 'stranded assets' represents a vital way to link the scientific necessity to phase out fossil fuels as part of a platform to address climate change with the financial sector that controls the flow of finance and determines the evolution of the global economy.

The concept is based on an understanding that there is a global scientific consensus about the causes of climate change, leading to agreement about the scale of CO_2 reductions necessary to keep planetary warming within the 2 °C limit or 1.5 °C agreed in Paris. These reductions mean that a significant proportion of known fossil fuel reserves have to remain in the ground and so fossil fuel companies will be required to write down the value of these assets since there is worldwide policy agreement that they cannot be exploited as previously planned, i.e. by being burned and thereby emitting carbon dioxide.

To dramatize the risks from not addressing asset stranding, campaigners also coined the term 'carbon bubble', to link the way urgent action on climate change could lead to financial instability if it were not followed by a rapid revaluation and adjustment of portfolios held by banks and on which the value of pensions and other investments depends (see the illustration in Fig. 1). In his introduction to a special issue on stranded assets,[1] Ben Caldecott defined the 'carbon bubble' as 'the hypothesis that unburnable carbon would mean that upstream fossil fuel assets were significantly overvalued, potentially creating a financial bubble with systemic implications for the global economy'.

An attempt was made to measure the extent of these 'stranded assets' in a landmark report from Carbon Tracker and Grantham Research Institute[2] published in 2013. They suggested that between 20 and 40% of assets booked as reserves by the world's corporations could not be realized if climate targets were to be met, and that for fossil fuels reserves this figure might rise as high as 60–80%. A study by the International Energy Agency from 2015 estimated that the value of assets at risk would range from \$4.2 trillion to \$43 trillion by 2100.[3] Carbon Tracker defined stranded assets as 'fossil fuel supply and generation resources which, at

[1] 'Stranded Assets and the Environment', special issue of *Journal of Sustainable Finance and Investment*, 7/1 (2017).

[2] Tracker, C. and Grantham Research Institute (2013), *Unburnable Carbon 2013: Wasted Capital and Stranded Assets* (London: Carbon Tracker).

[3] International Energy Agency (2015), *World Energy Outlook Special Briefing for COP21* (Paris: IEA).

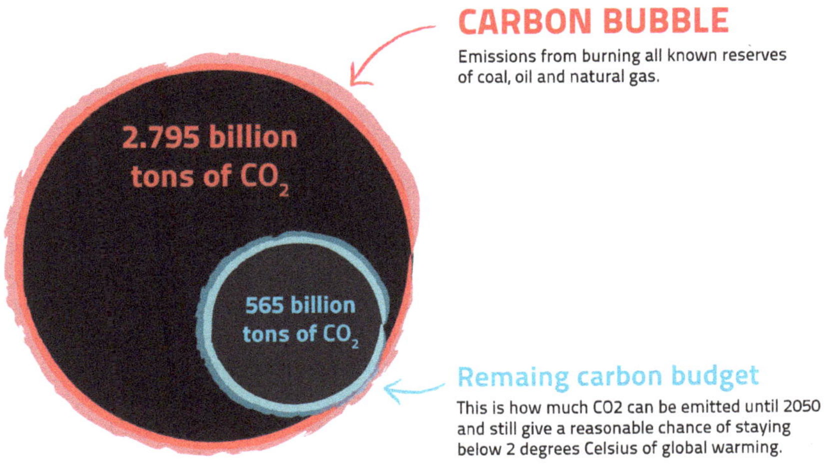

Fig. 1 The Carbon Bubble: graphical representation of the volume of unburnable fossil fuel reserves (*Source* Graphic by Felix Mueller based on data from the Carbon Tracker Initiative, 2013; thanks to Wikimedia Commons for making this graphic available free of charge [Available open source online here: https://commons.wikimedia.org/wiki/File:CarbonBubble_ENG.svg])

some time prior to the end of their economic life (as assumed at the investment decision point), are no longer able to earn an economic return (i.e. meet the company's internal rate of return), as a result of changes associated with the transition to a low-carbon economy'.

To give some academic grounding to the concept, Oxford University set up a Stranded Assets and Transition Finance Programme at its Smith School. They defined some of these risk factors that might be causing the stranded of assets:

- Environmental challenges (e.g. climate change, water constraints)
- Changing resource landscapes (e.g. shale gas, phosphate)
- New government regulations (e.g. carbon pricing, air pollution regulation)
- Falling clean technology costs (e.g. solar PV, onshore wind)

- Evolving social norms (e.g. fossil fuel divestment campaign) and consumer behaviour (e.g. certification schemes)
- Litigation and changing statutory interpretations (e.g. changes in the application of existing laws and legislation)

It's clear that some companies are going to be faced with the reality of these stranded assets sooner than others. The climate crisis is already impacting on insurance companies, as a growing number of extreme weather events and extreme temperature changes cause flooding, greater subsidence, and the failure of infrastructure that is now being pushed beyond its tolerance. So it is unsurprising that insurance companies have been some of the leaders in the field of sustainable finance.

The criticism often made of the financial sector is that it takes a very short-term perspective, which is an explanation for why action on climate has come so late. A sector that is a clear exception to this is the pensions industry, which has to ensure that its investments will pay a return to beneficiaries 40 or even 50 years from now. So actuaries and accountants valuing pensions are also very focused on the risk from stranded assets and the pensions industry has also taken a lead and lobbied and supported policy-makers on this agenda.

Although the concept of stranded assets has transformed the conversation on climate change and energized the sustainable finance agenda, there are two important caveats. First, we need to be clear that the sorts of assets covered by this term will continue, in themselves, to have economic value. So if you put petroleum into your car it will still run and coal will still heat your home or drive a turbine. Stranded assets become stranded because policy-makers choose to change the rules of the economic game, so the stranding is a result of political decision-making rather than economic necessity.

Secondly, although I have focused here primarily on climate and that is certainly the first area where the concept of stranded assets is driving the policy agenda, the issue is much wider and indeed includes what is now defined as ESG (Environmental, Social and Governance) impacts of economic activity. For example, the EU's High-Level Expert Group on Sustainable Finance[4] identified intensive agriculture as a sector that is,

[4] HLEG (2018), *Final Report 2018 by the High-Level Expert Group on Sustainable Finance* (Brussels: European Commission).

through its unsustainable practices, eroding its own productive base and stranding itself through loss of topsoil, contamination of water supplies, the collapse in numbers of pollinators, and so on.

As we introduce policies to limit and regulate the activities of a range of sectors that are profitable in the short term but are ultimately threatening not just future profits but the survival of the human species, some have raised concerns that stricter environmental standards are in themselves creating stranded assets. I agree and elsewhere I have explored how a changing legislative framework can diminish assets values, creating stranded assets that can then be removed from the portfolios of financial companies (see Fig. 3 in Chapter 4 for more detail):

> [An] area where legislation is likely to affect the value of assets is the extractive industries, as policy-makers enforce higher standards on mining companies, especially in terms of their activities in emerging markets.. From 1 January 2021 the new law – the Conflict Minerals Regulation – will come into force across the EU. It requires that EU importers of 3TG metals (tin, tungsten, tantalum and gold) must meet international responsible sourcing standards, set by the Organisation for Economic Cooperation and Development (OECD). From 2021, EU manufacturers of electronic goods using these four metals will have to monitor their supply chain to ensure they only import these minerals and metals from responsible and conflict-free sources. ... This is likely to have an impact on the value of shares in both extractive companies and manufacturers that rely heavily on these metals, especially mobile phone manufacturers. It may result in the stranding of the shares of companies that do not abide by the highest standards.

From the perspective of any individual company or sector this can seem unsettling, since pro-environment policy change can change the value of their assets, but that is really what management of the sustainability transition entails. And with sufficient time to plan and clear signalling, investors can anticipate legal changes and shift their investments accordingly.

4 Divesting from Fossil Fuels

So far we have explored how economists and financiers began to create a narrative linking the need to take action on the climate and environmental crises to the operation of and risks faced by financial markets. But the move towards sustainable finance has arisen from the grassroots as well as being developed in boardrooms and parliaments.

Much of the mobilization around the active removal of consumer or corporate finance from the fossil fuel industry has mobilized around the concept of 'divestment'. Like the concept stranded assets, this term began originally as a neutral term to mean the disposal of any asset but it has increasingly come to be associated with disinvestment from fossil fuel assets. Academics at Oxford's Smith School define divestment as follows: 'Divestment is a socially motivated activity of private wealth owners, either individuals or groups, such as university endowments, public pension funds, or their appointed asset managers. Owners can decide to withhold their capital—for example, by selling stock market-listed shares, private equities or debt—firms seen to be engaged in a reprehensible activity. Tobacco, munitions, corporations in apartheid South Africa, provision of adult services, and gaming have all been subject to divestment campaigns in the twentieth century'.

As Plantinga and Scholtens identify,[5] there is now a powerful and growing 'advocacy network that encourages institutional investors to divest from their holdings of fossil fuel stocks'; this is what we could describe as the divestment movement and it is mobilized under the banner 'Keep it in the ground'. The movement is already more than a decade old, with the 'Fossil Free Divestment Movement' being inaugurated in 2010: it claims credit for having shifted $14trn out of fossil fuels since then.

Much of the focus of the divestment movement has been on pension funds, both because they are such large global investors and also because they need to guarantee returns in the long term. Gupta et al. find that, across the world, pension funds have invested heavily in fossil fuels[6]:

> The UK's Local Government Pension Scheme had an estimated USD 14.9 billion and USD 19.8 billion invested in FFs in 2014 and 2017, respectively. In 2015 and 2016, the Dutch fund ABP held USD 9.2 billion and USD 11.4 billion in fossil investments... Colorado's Public Employee Retirement Association held over USD 1.5 billion in FF assets in 2018.

[5] Plantinga, A. and Scholtens, B. (2021), 'The Financial Impact of Fossil Fuel Divestment', *Climate Policy*, 21:1, 107–119.

[6] Gupta, J., Rempel, A. and Verrest, H. (2020), 'Access and Allocation: The Role of Large Shareholders and Investors in Leaving Fossil Fuels Underground', *International Environmental Agreements*, 20, 303–322.

The argument around this as a strategy to influence company activity revolves around the question of whether it is better to divest from shares or to hold on to them and engage in what is known as 'shareholder activism', i.e. to use the power of being a shareholder to try to influence a company's strategy and policies. If large public pension funds sell their assets they may be bought up by investors with fewer environmental scruples, or even shift the risk of stranded assets to the global South. However, it is questionable the extent to which a small number of dedicated shareholders can change the policies of these vast global corporations.

Scholars do agree, however, that divestment does not negatively impact returns to major investors. Plantinga and Scholtens find that 'screening out fossil fuel stocks has no significant impact on the return and the risk of a global well-diversified portfolio of industry indexes and that 'divestment from fossil fuel companies does not influence total financial risk for the investor'. They conclude that 'divestment will help change the mindset in the required direction of reducing the use of fossil fuels, and does not financially hurt investors and their beneficiaries'.

This is confirmed by research from Carbon Tracker,[7] which found that 'the value of share offerings in fossil fuel producing and related companies dropped by $123 billion in the last decade, underperforming a key world equities index by 52%'. Conversely, renewable energy stocks have outperformed the market average. This cannot all be considered the achievement of divestment campaigns, but the combination of changing government policy and active local campaigning is making fossil fuel investments unprofitable and it has now become a part of portfolio managers fiduciary duty to respond to this by divesting from fossil fuels.

And pension funds and other large-scale investors continue to face difficult decisions. Their holdings in fossil fuel assets are vast and are likely to be significantly overvalued as we move to tackle the climate crises, whether the bubble bursts suddenly or the loss in value is more orderly. In addition, the scale of capital at their disposal makes it difficult for them to find alternative investments, especially when renewable energy projects, for example, are often on a smaller scale.

[7] Carbon Tracker (2021), *A Tale of Two Share Issues: How Fossil Fuel Equity Offerings Are Losing Investors Billions* (London: Carbon Tracker).

5 Wider and More Radical Action on Climate Finance

Part of the purpose of the divestment campaign was always to shift the dial in terms of the narrative and sense of urgency around the climate crisis. As the proportion of carbon dioxide in the global atmosphere rose ineluctably while governments continued to support their fossil fuel industries, for some campaigners raising the issue of taking cash away from this dangerous sector was always part of a campaigning strategy to mobilize public opinion in favour of keeping fossil fuels in the ground.

There has been only limited study of the public policy impact of the divestment campaign but one interesting exception is an analysis of the 2019 *Divest Scotland* campaign, coordinated by Friends of the Earth Scotland and focused on members of the Scottish Parliament.[8] For a country that has benefited over decades from the North Sea oil and gas field, the aim of the campaign was to shift the opinion of Scottish politicians and to persuade them to commit not to invest their own pension fund in the fossil sector, nor to allow any fossil investments by the Scottish National Investment Bank. Muncie lists a number of interrelated objectives inherent in any divestment campaign: awareness raising, alliance building, and network creation. Drawing conclusions from his interviews with Scottish climate activists he found an 'uneasy swirl of motives oscillating between morality and markets' that lacked consistency:

> The simultaneous mobilisation around frames of climate justice (morality) and finance (markets) presents a significant challenge. On the one hand, a frame of climate justice demonstrates divestment's commitment to more radical aspirations and a holistic approach to tackling climate change, while on the other hand, a finance frame perpetuates notions of ecological modernisation and market solutions.

It is fair to conclude that the movement to shift finance out of fossil fuels attracts people with very different motivations and understanding of finance. But the political motivation is consistent and aligns with the statement by one of the founders of the movement,

[8] Muncie, E. (2020), 'Investing in Climate Solutions? An Exploration of the Discursive Power and Materiality of Fossil Fuel Divestment Campaigns in Scotland', *Journal of Environmental Studies and Sciences*. https://doi.org/10.1007/s13412-020-00653-2.

Bill McKibben, who said that 'money is the oxygen on which the fire of global warming burns' and 'if it's wrong to wreck the climate, it's wrong to profit from that wreckage'.[9]

Seyfang and Gilbert-Squires[10] analyse the campaign to push retail banking in a more sustainable direction in the UK. They identify the gap between the mainstream high-street banks (most of which have a globally investing commercial arm attached) and the ethical, pro-sustainability pioneers such as Triodos and the Ethical Building Society. There is really no comparison in terms of scope and scale, which is why the focus on strengthening sustainability requirements of banking regulation for all is essential in parallel to encouraging retail customers to move their individual accounts. In their conclusions the authors argue that 'the ideology of the current economic regime is pervasive throughout society, for example, economic education tends to be grounded in a neoliberal, rather than an ecological or new economics perspective… we find a critical point of intersection between [pioneering ethical banking] practices based on social and environmental return, and a capitalist system that values profiteering'. Their suggestion that some of these radical pro-sustainability banks might be involved in school education is an interesting one.

Other campaign groups have a bolder idea of how education might work. Extinction Rebellion has sent shockwaves through the environmental campaign community since it was launched in Trafalgar Square, London on 31 October 2018. The intention was to heat up the rhetoric and sense of urgency around the need for climate action—something that worked synergistically in the UK with the passage of motions declaring 'climate emergencies', beginning with Bristol in November 2018. The key targets of both was to focus back on staying within the 1.5 °C warming limit and achieving carbon neutrality by 2030 rather than the risky and unambitious date of 2050.

[9] McKibben, B. (2012) 'Global Warming's Terrifying New Math', *Rolling Stone*, 19 July. Available at: https://www.rollingstone.com/politics/politics-news/global-warmings-terrifying-new-math-188550/ (Accessed 28 August 2021).

[10] Seyfang, G. and Gilbert-Squires, A. (2019), 'Move Your Money? Sustainability Transitions in Regimes and Practices in the UK Retail Banking Sector' *Ecological Economics*, 156, 224–235. https://doi.org/10.1016/j.ecolecon.2018.09.014.

The obvious fact that the finance sector was not on course to achieve either of these objectives gave rise to the launch of 'Money Rebellion', which began in the spring of 2021 with the intention of 'rejecting the economic rules and financial institutions that are killing us'. Their stated aim is to highlight how banks are among the institutions which they claim are 'prioritising short-sighted, short-term profits over long-term survival' and have 'demonstrated they can be trusted with neither our money nor our lives'. They launched a wave of physical attacks on the infrastructure of the finance sector which they claimed was to highlight 'the deadly role of banks in what is a suicidal economic system that, by design, financially incentivises harm to biodiversity, the climate, and our future'.

At the time of writing (August 2021), the campaign group have sprayed fake oil over the front of the Bank of England, attacked the London HQ of Barclays Bank headquarters in London, and led protests in key financial centres across the world including New York, Paris, and Vancouver. As with the more mainstream divestment movement, part of the purpose of XR's Money Rebellion is to draw public attention to the way our own money is funding the ongoing climate and ecological emergencies, and especially the way public money is still being spent subsidizing the fossil fuel industry (Fig. 2).

In concluding this chapter it is important to recall again that climate is only the first, and most urgent, of the ecological and social crises that sustainable finance seeks to address. In Chapter 4 we will discuss what is known as the 'ESG agenda', namely the ways in which wider environmental impacts, as well as social and governance impacts, are all part of what must be considered when we define investments as sustainable. But before that, in the following chapter we will focus on some of the key concepts of definitions that help us to distinguish what is sustainable finance from what is not.

Fig. 2 Extinction rebellion demonstrate at the New York Stock Exchange, October 2019 (*Source* Photo by Felton Davis; thanks to Wikimedia Commons for making this graphic available free of charge [Available open source online here: https://commons.wikimedia.org/wiki/File:017_XR_at_Stock_Exchange_(48861040823).jpg])

CHAPTER 2

What Puts the Sustainable into Sustainable Finance

Abstract This chapter offers a basic introduction to the existing financial system: how does capital flow through the financial system and why does this create negative environmental and climate impacts? It explores who are the different actors and how they are motivated and where are the relevant points of potential change, with a special focus on the insurance industry. Market trends and how to interpret them are explained, followed by consideration of what we mean by a green bond and an account of the green bond market since its inception. Finally, we consider how to shift large-scale investment towards sustainable sectors by a combination of socially responsible investment and pro-sustainability investment indices.

Keywords Financial markets · Pension funds · Insurance · Green bond · Socially responsible investment · EU low-carbon benchmarks

1 What Do We Mean by Finance?

Finance is a closed book to most of us. A mysterious world where the initiates engage in dark arts that we don't understand but suspect are probably harmful. Understanding finance has enabled the UK to maintain its status as a global economic power as our manufacturing domination has waned. We are told that without it—the jobs it provides, the tax it

© The Author(s), under exclusive license to Springer Nature Switzerland AG 2022

M. Scott Cato, *Sustainable Finance*,
https://doi.org/10.1007/978-3-030-91578-0_2

Table 1 Key players in financial markets

Lenders	*Intermediaries*	*Markets*	*Borrowers*
Individuals	Banks	Interbank	Individuals
Companies	Insurance companies	Stock Exchange	Companies
	Pension funds	Money market	Central government
	Mutual funds	Bond market	Municipalities
		Foreign exchange	Public corporations

yields—we could not have the public services we prize. But we are told very little about how it works. So this first part of the chapter is mostly about defining terms and explaining the language of finance, before we go on to consider how we might decide what types of can legitimately be called sustainable.

An introductory textbook informs us that financial 'markets are all about the raising of capital and the matching of those who *want* capital (borrowers) with those who *have* it (lenders)'.[1] This is important because it tells you that the way finance is structured within a capitalist economy is all about the transfer of debt. Table 1 is taken from the same introductory textbook where it is labelled, 'the debt merry-go-round', which is as good a summary of financial markets as any. It traces the path that debt, or finance, takes from those who lend it to those who borrow it. The image of a merry-go-round is a good one because it conveys the sense that it is the *circulation* of money that is crucial to making the system work.

There is a common misconception that the loans balance the deposits and that deposits precede loans, but this is not the case. In fact, for the vast majority of finance in circulation in the global economy today, the creation of the debt brought the money into existence. So why do we trust it? It is because belief in the lender is so fundamental to this financial system that we refer to the ability to borrow as 'credit', from the Latin *credere*, 'to believe'.

[1] Valdez, Stephen and Molyneux, Philip (2016), *An Introduction to Global Financial Markets* (Palgrave Macmillan).

A study of the financial system is endlessly fascinating but for the purposes of this book we are going to focus particularly on a few of these players. Banks matter, because they are the heart of the system and the source of much of the money in circulation, so they have their own Chapter 6. Individuals include you and me, although we are more likely to have money invested via the intermediaries, especially insurance companies and pension funds, unless you happen to be very wealthy. For us, what's important in terms of achieving sustainability is that we know the impact of the investment decisions being made on our behalf by these intermediaries. This is the question of 'non-financial reporting' and 'mandatory disclosure' that we will come onto in Chapter 5. In the rest of this chapter we will consider the bond market and the growth of green bonds and the role of the vast amount of money in pensions and insurance and its role in sustainable investment. The following section provides you with some of the language and concepts you'll need for the discussions that follow.

2 More Key Questions and Definitions

We will delve into this creation of money in a little more detail when we discuss banking in Chapter 6. For this chapter it is sufficient for us to ask a few important questions, once we know that money is being created through the creation of debt. Who has the power to call money into existence, while simultaneously creating a parallel debt, and what conditions are there on that power?

This question became very important during the 2008/2009 financial crisis when citizens realized not only that the existing regulations had allowed far too much risk of default, but that they owned that risk! Nor understanding this proved costly to citizens of countries with large financial sectors like the UK, the US, and even Ireland. Failing to impose conditions on finance with regard to sustainability as well as stability could prove even more costly to the planet as a whole. Because finance is the fuel of the global economy and we need it to flow towards sustainable activities like wind-power and away from destructive sectors like fossil fuels if we are to have any chance of resolving the climate and ecological crises we are facing.

If we start from the understanding that finance is about the movement of debt from one party to another we need to ask some key questions about that debt. Obviously the first one is: how large is it? If you're borrowing a few thousand to insulate your house or to buy new equipment for your business you would raise it via a retail bank or mortgage company. If you were a large corporation you might issue the debt as a bond (see below) or as (equity) ditto.

Exactly how you raise that debt might depend on how long you wanted to borrow if for, how likely you were to pay it back, and how much you would be charged to borrow it (the interest rate). These three are obviously interconnected. Governments can borrow more cheaply than anybody else because they are least likely to default and they borrow over the very long term (obviously this rule doesn't apply in the case of governments like those of Argentina that routinely default). If you have a very risky venture in mind the risk premium will be reflected in the interest rate—banks or venture capitalist might be prepared to lend you the money but only if they are tempted by the chance of sufficient returns.

Some of the largest pools of cash in the financial markets are pension funds and insurance companies. Because they are holding money throughout people's lives they have extremely long time horizons. This explains why they have been some of the most active in terms of the sustainable finance agenda because the climate crisis is greatly increasing the risks they face.

To finish this initial introduction to finance we need to distinguish between three ways companies can raised finance.

The first and most obvious is a *loan*. A loan means that a financial institution—usually a bank—will advance you money for a fixed period of time (or 'term') and at an agreed cost (the 'interest rate').

Secondly, we have a *bond*, which also means borrowing money but in a more structured way. A bond is a piece of paper or digital equivalent stating the value of the money advanced and the terms of which it will be paid back, for example, to be paid back after ten years with two instalments of interest at 7% paid twice a year. A range of different organizations can issue bonds, including the national government, local government, and private companies.

Financial commentators often distinguish between 'debt' and 'equity' as means for funding a new venture. Entrepreneurs raised *equity* finance through selling *shares* in their company, thus sacrificing control in return for funds. The first time this happens it is called a 'new issue' and when companies issue shares later in their development this is called a 'rights issue'. In return for buying the shares investors receive income in the form of a dividend and potentially capital gains if the shares increase in value. These returns are not guaranteed, though, making this a riskier form of investment. Somewhat confusingly, owning a proportion of a company is also sometimes called owning *stock* which means owning an unspecified number of shares.

For each of these kinds of finance, financial actors with an awareness of the climate and ecological emergencies can bring their preferences to bear. For example, they can choose to invest only in companies that they believe have a future in the sustainable economy of 2030—so renewable energy rather than fossil fuels. Or they can limit their share purchases to companies that perform well on sustainability criteria, or use their shareholder power to pressure for that (shareholder activism). Companies may choose to issue special bonds—'green bonds'—that ensure the money is invested in a sustainable way. In the case of the lenders, they can choose to act like many of the German public banks in lending only to companies that have strong sustainability credentials—of which more in Chapter 6.

3 Insuring and Evaluating the Risks

Who are the key sectors that are shifting money towards sustainable sectors? When I first started engaging in sustainable finance from the policy side, I was surprised to encounter top-level executives from insurance companies not only keen to speak with me but actually speaking the same language. Since they have the ability to shift billions of finance across the world we need to delve a little deeper into how they work and the role they can play in accelerating the sustainability transition. Not only do they control large funds, they also have a long-term perspective:

> Where pension funds exist, these funds of money, along with those of insurance companies, are key determinants of movements in the markets. They have to look ahead to long-term liabilities, and will assist the borrowers of capital by buying government bonds, corporate bonds, corporate equities and so on.[2]

Mark Carney, at that time governor of the Bank of England, helped to theorize this long-term approach to financial returns and its connection

[2] Valdez, Stephen and Molyneux, Philip (2016), *An Introduction to Global Financial Markets* (Palgrave Macmillan), p. 171.

with climate change in a speech called 'the Tragedy of the Horizons', a speech he gave at Lloyds, the heart of the global insurance industry, in 2015.[3] During the speech he claimed that 'insurers are amongst the most determined advocates for tackling [climate change] sooner rather than later' and questioned why more action was not being taken to tackle climate change. He compared this failure to the famous economics conundrum of the 'tragedy of the commons'—a suggestion that when property rights are not clearly defined environmental crisis can result. He coined the new term 'tragedy of the horizon', arguing that business, political, and regulatory cycles were too short to encompass the need for longer-term vision and action on climate. He outlined ways in which insurers could respond to the multiple risks to their business—and to society—posed by this inability to plan for the long term in a speech that is considered to have increased focus on sustainable finance across the world.

Insurance is a form of finance that is seeking long-term security rather than short-term rewards: 'people take out a life policy and perhaps do not change this for the whole of their life. As a result, the business is stable and can invest for the long-term. The nature of this business is such that it is natural for them also to invest funds to provide pensions in the longer-term' (ibid. 373).

When you think about what they do, it becomes obvious why pensions and insurance providers are leading the sustainable finance discussion.

Much of the activity in financial markets has an incredibly short time-horizon, indeed high-velocity trading is essentially split-second gambling by computers without any human interaction. This has been seen as the dynamic and exciting part of the financial sector, and it is certainly lucrative. Pensions and insurance companies operate at the other extreme. They are seeking low-risk, long-term, stable returns so that they can pay people's pensions in 30, 40, or 50 years' time.

Their long time-horizon gives them a reason to take sustainability seriously. As an obvious example, if we are aiming for net-zero by 2050 that means any investments made in fossil fuel companies will not pay returns to pension holders after that date. Less obviously, but also of key importance, if we continue to erode and undermine our soils with excessive use of pesticides, or lose essential pollinators, or if some prime farming areas

[3] Carney, Mark (2015), 'Breaking the Tragedy of the Horizon—Climate Change and Financial Stability', Speech at Lloyd's of London, London, 29 September 2015.

become arid and no longer productive, investments made in agribusiness in those areas will not yield a return. This explains why some of the world's leading pensions providers are also key advocates for sustainable finance.

Because many pensions include life insurance the pensions and insurance industries are closely linked and many of the larger providers offer both kinds of policies. And both face issues with the costs of climate change. Insurance premiums are rising as extreme weather events provoke flooding, more severe hurricanes, droughts, forest fires, and a range of other calamities that destroy insured property as well as lives.

It is difficult to make a reliable estimate of the increasing costs of climate damage, especially because it arises from a vast array of different climatic disasters ranging from more and more intense wildfires caused by heatwaves, stronger tropical storms resulting in catastrophe, damage to property, and loss of life especially in countries with less ability to invest in resilience and protection measures; rising sea-levels causing flooding especially in low-lying islands and countries such as Bangladesh; and additional energy in the atmosphere leading to extreme weather events including flash flooding. Hand in hand with denial, some of it stoked by the very fossil fuel companies that have been driving us to disaster, the systemic complexity of climate change has helped to conceal the vast economic costs until it may be too late.

However, estimates have been made for the cost of individual climate-related events. The average estimate of damages from Hurricane Harvey, which struck Texas and Louisiana in August 2017 has been assessed at about US$90bn, acknowledging that this includes only direct economic costs US$67bn. and excludes deaths, injuries, and dislocation or displacement of citizens.[4]

The European Environment Agency provides regular updates on the economic losses from climate-related extremes, although does not attempt to assign these to anthropogenic climate change.[5] Their estimates are that between 1980 and 2019, the losses amounted to €446bn in the EEA member countries. An earlier version of the report notes that these

[4] Frame, D. J., Wehner, M. F., Noy, I., et al. (2020), 'The Economic Costs of Hurricane Harvey Attributable to Climate Change', *Climatic Change* 160: 271–281. https://doi.org/10.1007/s10584-020-02692-8.

[5] https://www.eea.europa.eu/data-and-maps/indicators/direct-losses-from-weather-disasters-4/assessment.

figures are inflated by a small number of extreme and very costly events, especially 'the 2002 flood in Central Europe (over EUR 21 billion), the 2003 drought and heat wave (almost EUR 15 billion), and the 1999 winter storm Lothar and October 2000 flood in Italy and France (both EUR 13 billion)'.[6]

Tropical storms in the Pacific are likely to be much more costly on a human scale, with so many low-lying islands and countries with less ability to invest in prevention measures. The Asian Development Bank estimated that losses from typhoons and earthquakes cost the Philippines around $1.6bn each year.[7] The constant onslaught from extreme weather also has a significant effect on the country's economic activity, with effort displaced to repair damage and whole islands and communities finding their economic infrastructure destroyed. The most devastating recent typhoon was Haiyan (called Yolanda in the Philippines) that struck in 2013 and was estimated to have cost $5.8 billion.[8] The human costs were even more devastating with 7,000 people killed, nearly 2 million left homeless, and more than 6 million displaced.

As discussed in the following chapter, it is clear that the loss and damage resulting from climate change is mostly impacting on the Global South. But conversely, and shockingly, the costs in those communities are often underestimated because of their not being able to afford to insure against losses and because costing of lives is based on earning potential, which is lower in countries with less economic power. (I can't let this pass without express my moral outrage.) This gives an idea of how the climate crisis reveals a series of interconnected injustices that dominate our global economy.

The New Zealand Treasury Department has boldly produced an estimate on the likely cost of the impacts of man-made climate change in their country between 2007 and 2017.[9] They provide estimates for the cost

[6] https://www.eea.europa.eu/data-and-maps/indicators/direct-losses-from-weather-disasters-3/assessment-2.

[7] Strobl, E. (2019), 'The Impact of Typhoons on Economic Activity in the Philippines: Evidence from Nightlight Intensity', Asian Development Bank Economics Working Paper Series, No. 589. https://www.adb.org/sites/default/files/publication/515536/ewp-589-impact-typhoons-philippines.pdf.

[8] https://www.bbc.co.uk/bitesize/guides/z9whg82/revision/4.

[9] Frame, D., Rosier, S., Carey-Smith, T., Harrington, L., Dean, S., and Noy, I. (2018), *Estimating Financial Costs of Climate Change in New Zealand* (Wellington: NZ

of insured damages associated with flood attributable to anthropogenic influence on climate are 'currently somewhere in the vicinity of' $120M and $720M for economic losses associated with droughts'. They note that the absence of studies on costs associated with 'storm damage, hailstorms, wildfire, frosts or tornadoes' means that they have omitted at least NZ$279M in weather-related losses during that decade from their estimates.

Suffice to say that these estimates are inaccurate and inadequate but they give a sense of the massive and costly destruction we have in store if we do not take urgent steps to restructure our economic system.

4 Hey Big Spender

The insurance industry has been playing a leadership role on the sustainable finance agenda since 2012, when the UN Environment Programme Finance Initiative launched its Principles for Sustainable Insurance. Its most recent report[10] restates the centrality of the insurance industry to the process of bearing the risks of the climate crisis and having the financial heft to accelerate climate action:

The insurance industry is one of the largest global industries with more than USD 6 trillion in world premium volume and US$36trn in assets under management. As such, insurers hold a significant portion of global economic assets and liabilities on their balance sheets. As risk managers, insurers and investors, the insurance industry can play a leadership role in building climate-resilient communities and in accelerating the transition to a net-zero emissions economy.

As long ago as 2007, leading insurers formed the organization ClimateWise, a global industry body that assesses its members' progress to promote climate awareness among customers and reduce the environmental impact of their business. The Association of British Insurers takes forward its work under three headings: Adapt (risk mitigation), Innovation (investing in pro-sustainability technologies), and Invest (reducing

Treasury). https://www.treasury.govt.nz/sites/default/files/2018-08/LSF-estimating-financial-cost-of-climate-change-in-nz.pdf.

[10] UNEP (2021), 'Insuring the Climate Transition: Enhancing the Insurance Industry's Assessment of Climate Change Futures'. https://www.unepfi.org/psi/wp-content/uploads/2021/01/PSI-TCFD-final-report.pdf.

the amount of the portfolio in assets that may become stranded while prioritizing investments that accelerate the sustainability transition).

To gain an insight into how those inside the insurance industry perceive the risks posed to their business by the climate crisis we could do worse than consult a report on this topic from the global management consultancy McKinsey, published in 2019.[11] While the report is couched in terms of 'threats and opportunities' it does not pull its punches in terms of how the rapidly changing climate is causing equally rapid changes to what has traditionally been a conservative sector. The report is clear that insurance will need to reshape its business models and that this is overdue: 'Some others have publicly committed to reducing their exposure to carbon-intensive industries by 2030 or 2040. In recent interactions with industry executives, more than half have said that the industry's response so far has been underwhelming and inadequate—even though the vast majority said that responding to climate risk is either "very important" or "a top priority"'.

The advice McKinsey offers to its clients for changing their business models indicates exactly how the insurance industry is now an ally of those campaigning hard to shift finance in the direction of pro-sustainability investment. They suggest five actions insurers need to take:

- Stress-test total exposure against projected climate hazards
- Build resilience and rebalance portfolios
- Help organizations mitigate climate risk
- Create innovative products to address climate-related risk
- Revise investment strategies

Stress-testing is essential, meaning that the company understand the size of its exposure which is certain to be underestimated as climate impacts accelerate. McKinsey redefines insurers as 'traders of climate risk' and suggests that they need to lead understanding of the impacts of the climate crisis. Helping organizations mitigate risks is nothing new for insurance companies, who have always advised homeowners on property security, for example. But they are now being encouraged by insurers to 'work with the public sector to improve building standards and policies;

[11] 'Climate Change and P&C Insurance: The Threat and Opportunity', McKinsey (2020). https://www.mckinsey.com/industries/financial-services/our-insights/climate-change-and-p-and-c-insurance-the-threat-and-opportunity#.

an analysis of risk models may suggest limits to building in flood-prone areas, for example', something environmental campaigners have long argued for. The innovative products they suggest might include products that share risks between insurers, those in need of insurance, and those whose activities might increase that risk.

The other side of the insurance business is how they invest the premiums paid to them. As already identified, these are investments for the long term and so the insurance sectors are leading on the shift in investments that will foster the sustainability transition. McKinsey advises: 'Insurers should reevaluate their investment-allocation strategies as the economy transitions toward long-term decarbonization, which may cause rapid asset repricing and portfolio volatility, particularly for carbon-intensive investments'. The final paragraph of the report indicates how the sustainability-focused regulation in the financial sector (discussed further in Chapter 4) is impacting on the perceptions and strategies of the insurance industry:

Insurers should also consider the environmental impact of their investments, just as banks and asset managers are doing, and follow a plan to shift significant portions of their portfolios toward supporting a sustainable, decarbonized economy. In addition to long-term benefits, this shift will help insurers demonstrate their proactive compliance, as regulators may enact policies and incentives that can affect the investments financial-services companies make toward their environmental, social, and governance (ESG) footprint.

5 My Name Is Bond, Green Bond

As we say above, issuing a bond is a way of companies borrowing money directly rather than through a bank loan. In recent years, bonds focused on pro-sustainability activities and companies have come to the fore but how can investors be sure about the credentials of the bond they think is positive for the environment or the climate?

The International Capital Markets Association laid down some voluntary principles for those issuing green bonds in 2018.[12] These are:

[12] ICMA, 'Green Bond Principles: Voluntary Process Guidelines for Issuing Green Bonds', (2018). https://www.icmagroup.org/assets/documents/Regulatory/Green-Bonds/Green-Bonds-Principles-June-2018-270520.pdf.

- Use of Proceeds: These must provide clear environmental benefits, preferably quantified by the issuer. Appropriate sectors include renewable energy; energy efficiency; pollution prevention and control; sustainable land management; biodiversity conservation; sustainable transport; climate change adaptation; circular economy production; and green buildings.
- Process for Project Evaluation and Selection: the bond issuer is required to make the environmental sustainability objectives clear to investors, and how these are measured.
- Management of Proceeds: to avoid mission slippage green bond proceeds need to be managed in a separate portfolio from the management company's other accounts.
- Reporting: issuers need to provide relevant and timely information about the impact of their investments to investors.

The green bond market has experienced exponential growth since the first green bond was issued by the European Investment Bank in 2007 (see Fig. 1). By December 2020 it reached an important milestone when the total value of green bonds issued reached US$1trn. November 2013 marked another important milestone in terms of credibility when the first corporate green bond was issued by Swedish property company Vasakronan. Other large corporate issuers include SNCF, Berlin Hyp, Apple, Engie, ICBC, and Credit Agricole.[13]

Green bonds are surging in popularity since the Covid-19 pandemic with the issuance of green, social, sustainable, and sustainability-linked bonds doubling in the first half of 2021. According to Bloomberg, the value of green bonds issued during that period represented a value of $248bn moving into sustainability sectors. Bonds issued to finance green and social projects increased from $71 billion throughout 2020 to $90.4 billion in the first half of this year. A new category of bond that includes sustainability performance targets that are required to pay more to holders if they are not reached is also expanding rapidly.[14]

There has been considerable concern about the absence of any independent scientific monitoring of the real impact of money invested in

[13] Climate Bonds Initiative: https://www.climatebonds.net/market/explaining-green-bonds.

[14] https://www.greenbiz.com/article/green-bonds-are-beating-all-expectations-post-pandemic-recovery.

Fig. 1 Exponential rise of the green bond (Figure includes Green bonds, loans, Sukuk [Islamic finance certificates] and green asset-backed securities. *Sources* Climate Bonds Initiative; Author's graphic redrawn by Angela Mak)

what are claimed to be 'green bonds' and claims of greenwashing have been widespread. As we will see in Chapter 4, the EU is working to establish global standards for green bonds and what should be excluded is as important as positive investment, as the example of fossil fuel companies claiming to be issuing 'green bonds' makes clear.

In May 2017, Spanish FF corporate Repsol raised €500 in a bond issue, claiming this would help them reduce climate emissions. As Saida Eggerstedt, head of sustainable credit at global investment firm Schroders commented, 'The Repsol green bond should remind investors how important it is we ask questions around what constitutes a valid target or key performance indicator... Investors must undertake due diligence and not just believe issuers' promises that the money they get from a green or other ESG bond is used solely for appropriate projects'.

In late 2020, the *Financial Times* interviewed a number of market players to explore their experience of greenwashing of green bonds.[15] Chris Bowie, a portfolio manager at TwentyFour Asset Management, pointed out that buying a green bond from a company whose main business is destructive is essentially pointless in terms of achieving sustainability benefits. Tom Chinery, a corporate bond portfolio manager at Aviva Investors, pointed out that sustainable investors should focus on the company as a whole rather than one particular aspect of their business they claim they will use the additional cash flow for. As a particularly egregious example, a 'green bond' issued by the Queensland state government in Australia was claimed to be invested in preserving the Great Barrier Reef while the state continued to support its massive coal industry.

The solution to greenwashing is better reporting and clearer measurement and definition of pro-sustainability impacts, a finding that is reinforced by an academic study of a survey of European asset managers.[16] They found that those most likely to be purchasing green bonds favoured bonds that are issued by non-financial corporates in industrial, automotive, and utility sectors, as well as by national governments. As well as the price, fund managers are looking for strong green credentials and weak reporting or lack of clarity are the main obstacles to growth in this market.

An academic analysis of corporate green bonds explores three possible motives issuers might have: signalling environmental commitment for PR purposes; a form of greenwashing with a similar intent, but no authentic commitment; using the fact that green investors are willing to accept lower returns to access a cheap form of financing.[17] The author concludes that the stock market responds positively to the issue of green bonds, especially when they are validated by an independent body. She also finds that companies do improve their environmental performance after issuing

[15] 'Investors Probe ESG Credentials of Bond Sellers on "Greenwashing" Fear', *FT*, 28 October 2020. https://www.ft.com/content/1bcbad16-f69e-47db-82fa-0419d674bb53.

[16] Sangiorgi, I. and Schopohl, L. (2021), 'Why Do Institutional Investors Buy Green Bonds: Evidence from a Survey of European Asset Managers', *International Review of Financial Analysis*, 75 (May), 101738. https://doi.org/10.1016/j.irfa.2021.101738.

[17] Flammer, C. (2021), 'Corporate Green Bonds', *Journal of Financial Economics* (available online, January). https://doi.org/10.1016/j.jfineco.2021.01.010.

a green bond, thus supporting the signalling thesis and undermining the greenwashing or cheap-acccess-to-capital theses.

Concerns about greenwashing persist and threaten the credibility of the whole market for sustainable finance. When investors who are seeking sustainable options see their money diverted towards fossil fuels or invested in companies or government that are simultaneously funding environmentally destructive activity it is clear that the green bond market is at serious risk of losing all credibility unless high, measurable and enforceable standards are applied. This concern lies behind the EU's decision to develop a sustainable taxonomy, or system of definitions, that can form the basis for applying clear standards and an EU kitemark for genuinely sustainable financial products (see Chapter 4).

6 Where There's Muck, There's Brass—The Relative Performance of Green and Other Investments

This chapter's heading sums up a preconception—even a false prejudice—about sustainable investments, namely that they represent philanthropic activity rather than being driven by the profit motive. This is summed up by the explanation of this proverb that dates from the early days of the industrial revolution that was driven Britain's northern cities that paid the price in terms of pollution and poor health: 'dirty or unpleasant activities are also lucrative'. In this section, we'll explore evidence that demonstrates that, at least in the case of sustainable investments, the data does not bear this out.

As discussed above, the shift in investments began as a moral or political crusade and under headings such as 'ethical investment', 'socially responsible investment', or sometimes 'impact investing'. It has a long history, almost as long as the history of capitalism itself. In the early days of trading in company shares Quakers set an example by refusing to invest in the slave trade while Methodists avoided investing in gambling and industries that used toxic materials. The strongly moral or religious motivation of SRI continued with religious leaders advising their congregations to invest in a way that would support the civil rights movement or starve the Vietnam War and apartheid South Africa of funds. Withdrawal of investment in tobacco has also featured strongly in the movement.

From the early 1990s, financial advisors began offering socially responsible indexes, meaning that they would screen companies before including them in indexes that investors could then choose. These have struggled to maintain credibility, leading regulators to strengthen reporting standards.

There are two alternative approaches: negative screening (meaning avoiding companies whose business creates harm, such as cigarettes) and positive investing (choosing companies that bring positive benefits, such as renewable energy). As we'll see in Chapter 5, as responsible investing has moved into the mainstream it has become subject to regulation in terms of both non-financial reporting and now full ESG (environmental, social, and governance) reporting.

Having been driven by conscience and an interest for only a minority of investors, responsible investing is now moving rapidly towards the mainstream so that by the beginning of 2016, assets managed in a socially responsible way amount to a global value of $22.89trn, representing 26% of all professionally managed assets, a proportion that had increased by 25% since 2014.[18]

While those motivated to invest in this way represented a minority of investors, their portfolios have enabled research to compare the performance of stocks scoring high on environmental, social, and governance performance with average or even 'mucky' stocks, thus testing the proverb.

Finnish student Ida Vehviläinen actually explores this question directly in her thesis which is called 'Does it actually pay off to be bad rather than good? Sin stocks, socially responsible investing, and the EU taxonomy'.[19] Her analysis also benefits from the close specification of what lies within the scope of a 'good' investment by using the EU sustainable finance taxonomy as its analytical frame (see more in Chapter 4). She constructed six portfolios using a range of stocks included in the STOXX Europe 600 index between 2003 and 2019. She found that 'sin stocks' do not provide statistically significantly higher returns than socially responsible stocks and

[18] Martini, A. (2021), 'Socially Responsible Investing: From the Ethical Origins to the Sustainable Development Framework of the European Union', *Environ Dev Sustain*. https://doi.org/10.1007/s10668-021-01375-3.

[19] Masters Thesis at the University of Vaasa, Finland: https://osuva.uwasa.fi/handle/10024/12535.

that excluding stocks that fall within the scope of the EU taxonomy negatively effects long-run returns. So avoiding the muck brings more brass rather than less.

Although the socially responsible investment movement was driven by ideological, political, or moral commitment, research has repeatedly shown that this may go hand in hand with strong financial returns.[20] Li and colleagues created a portfolio optimization model incorporating performance scores on E, S, and G criteria which they then applied to the performance of US stocks between 2005 and 2017.[21] They found that the SRI portfolio was able to simultaneously achieve investment returns and social value and may have outperformed traditional investment strategies.

Omura and colleagues found that these returns are also resilient during a global health crisis, by examing the performance of SRI/ESG investments against conventional investments during the COVID-19 pandemic. They found that responsible investments outperformed conventional investments to a greater extent during COVID-19; that responsible investment factors affected returns more during the pandemic; and that responsible investment was more resilient during the pandemic than conventional investment.

Finally, we can look for hard facts about whether the sustainable finance agenda has had the most important effect we would expect to see if it were effective, namely a decline in the value of stocks in companies that will not survive the sustainability transition, especially fossil fuel stocks. If the regulatory and reporting changes driven by the sustainable finance agenda are being successful, then we would expect the value of these stocks to fall. And, as we will see in Chapter 4, the objective of policymakers is that they should fall in an orderly but rapid way, rather than the pressure of stranded assets leading to the bursting of the 'carbon bubble' (see Chapter 1).

A report from spring 2021 by the London-based sustainable finance monitor Carbon Tracker revealed that the loss in value of such stocks is

[20] Omura, A., Roca, E., and Nakai, M. (2020), 'Does Responsible Investing Pay During Economic Downturns: Evidence from the COVID-19 Pandemic', *Finance Research Letters*, 101914, 31 December. https://doi.org/10.1016/j.frl.2020.101914.

[21] Li, C., Liepei, Zhang, Jun, Huang, Helu, X., and Zhongbao, Z. (2021), 'Social Responsibility Portfolio Optimization Incorporating ESG Criteria', *Journal of Management Science and Engineering*, 6:1, 75–85.

happening and accelerating. Researchers analysed the value of equities in fossil fuel producers together with fossil fuel dependent utilities, pipelines, and service companies. They compared their performance with issuance and returns to stocks in electric utilities and renewables/cleantech companies, as well as the general equity market measured by the MSCI (Morgan Stanley Capital International) index. The key findings of the report are reported in Table 1. In launching the report, Henrik Jeppesen, one of its authors, said: 'Investors have woken up to the fact that fossil fuel companies are no longer the growth stories they once were. Climate risk is now very much a material one that cannot be ignored and clean energy stocks are rapidly replacing the old order as the choice investment for a transitioning world'.

As the data in Table 2 shows, a comparison of the market performance of fossil fuel and related companies with an index of all companies shows that the former are underperforming by 70%. However, as we discovered

Table 2 Indicators of declining value of and confidence in fossil fuel stock

Indicator	*Metric*
Performance of energy stocks	Underperformed MSCI's All Country World Index (ACWI) by 70%
Proportionate size of issuance of FF vs. renewable/clean-tech	Nearly $640bn for FF compared with only $56 billion from renewables/cleantech companies
Appetite for FF stock	Accounted for 12% of total equity issuance proceeds in 2012, falling to 8% between 2014 and 2016 and to less than 1% in 2020, with the number of annual completed transactions falling by 75%
Loss and underperformance of FF stock	$123 billion loss in value; underperformed ACWI by 52%
Gains and performance of renewable/clean-tech	Outperformed ACWI by almost 54% and gained $77bn in value
Proportion of equity issue to existing/new holders	Secondary share sales have risen from 6% of equity proceeds in 2016 to 78% in 2019 (excl. Aramco IPO) and by 58% in 2020*

*Since 2016

Note Data relates to the decade between 2011 and 2021

MSCI: Morgan Stanley Capital International; ACWI: All Country World Index

Source Carbon Tracker (2021), *A Tale of Two Share Issues*, https://carbontracker.org/reports/a-tale-of-two-share-issues/

in the case of green bonds, when the FF sector is compared with renewable and clean-tech companies we see that the latter is still a minority of the market which, while it is expanding rapidly it is still a marginal player, represents only a tenth the overall value of fossil fuel stocks. The decline in new issues of equity is particularly revealing, suggesting that existing shareholders are buying new stock to protect their existing investments but new investors are not tempted by the risk that holding fossil fuel stock now represents.

Bloomberg reports a similar, and accelerating, trend, claiming that private equity 'is pouring capital into fast-growing sectors such as solar, carbon capture, and battery storage'. They report that by the middle of 2021 renewable energy investments had outstripped fossil fuel assets by a factor of some 25, so 25 times as much money has gone into renewables as into fossil fuels. This bodes very well for the sustainability transition.

With climate change at the top of the political agenda across the world, it is not surprising that fossil fuel stocks are under pressure. But with the reach and scope of non-financial reporting increasing every year (see Chapters 4 and 5) the number of unsustainable sectors that will struggle to return dividends and see the value of their stocks fall is likely to expand.

7 Shifting the Big Bucks

We often hear that we are all responsible for the environmental crisis because our investments in pension and insurance funds are actually being used in ways we never imagined. What this points to is the fact that very few of us actively invest our own money. We rely on investment brokers to do this.

But here's the thing: most funds are invested not according to the insight of the brokers themselves but according to indices that are regulated and monitored. An index is simply a collection of tradable shares gathered according to some criteria. For the indices most of us have heard of—the FTSE (Financial Times Stock Exchange), NIKKEI, or Dow Jones, for example—the criterion is only size. The FTSE100—familiarly known as the 'Footsie'—was created in January 1984 and calculates the moving value of the shares of the largest 100 companies registered to the London stock exchange. By contrast, the S&P (Standard and Poors) Global 100 includes stocks from the 100 largest global companies.

Keeping a track of the movement of so many shares can be very time-consuming and for this reason many of those who invest our money are what is called 'passive investors', which means they do a minimal amount of buying and selling and tend to follow indexes chosen by market analysts rather than making their own decisions. In effect, they are tracking the movements in the market and assuming that stocks are priced correctly to reflect the earnings of their parent companies.

Parallel to the arrival of ethical investment, investors began to assess the need to look inside the black box of the index and impose some criteria beyond earnings. This led to the establishment of indices linked to ethical or sustainability performance. An early example was FTSE4Good, an index established in 2001 which included only stocks that met a series of corporate social responsibility criteria. For example, the index will not include companies in arms production, tobacco, or coal and companies included in it will have met minimum standards human rights and governance. Hence people investing their money following this index will know that they are spared the worst of negative impacts and that their money is not doing harm in the world.

While an MEP I worked to develop the EU legislative framework for benchmarks to govern the use of such passive investing. The concept of these 'climate benchmarks' was to introduce two benchmarks to govern indices that would only include companies that would be beneficial in terms of the climate crisis.

Following tortuous political negotiations we arrived at definitions for two benchmarks:

- An 'EU Climate Transition Benchmark', where the underlying assets are selected, weighted, or excluded so that the resulting benchmark portfolio is on a decarbonization trajectory that can be scientifically verified and is regularly monitored.
- An 'EU Paris-aligned Benchmark', where the underlying assets are selected so that the resulting benchmark portfolio's GHG emissions are aligned with the long-term global warming target of the Paris Climate Agreement and are also constructed in accordance with the minimum standards on ESG laid down in the delegated acts.

So the latter benchmark is more stringent in terms of including the companies that are already Paris-compliant while the former includes companies who are not yet compliant but can prove that they are making a rapid and significant transition towards compliance. Both suffer from a significant weakness in that they do not exclude certain activities or sectors, even fossil fuel companies. The very weak standard of a 'do not significant harm' exclusion is the most we could achieve when negotiating with MEPs from right-wing parties.

Although the motivation here was good, the final legislation fell far short of the ambition because of lobbying from corporations brought into the political negotiations by mainstream politicians. I can promise you I talked myself hoarse trying to explain to my fellow negotiators that, if you water down the criteria until they no longer have imposed any meaningful restrictions on company behaviour, the benchmark becomes futile, but with limited success. However, for a major financial regulator to establish criteria for benchmarks that achieve high performance on sustainability criteria is a novel and important step towards enabling passive investors to guide their clients' money towards doing good in the world, rather than the reverse.

CHAPTER 3

The Chequered History of Climate Finance

Abstract Because it is the largest and most urgent crisis facing humanity, it is appropriate that climate change—and how to finance our response—has its own chapter in a book on sustainable finance. Here we consider what climate justice can tell us about how the balance of investment should be shared between the countries that have exploited fossil resources to grow rich and those that have fewer resources but are suffering more immediate and more severe impacts from the climate crisis. This, known as 'the loss-and-damage agenda' in UN negotiations, has been the source of repeated conflict at the UNFCCC process. The chapter also considers and compares climate investment packages from a range of the world's largest economies, known variously as Green New Deal or green stimulus. I then explore the link between colonial history and the need for reparations and routes to funding the loss-and-damage agenda. And I conclude by considering more radical proposals for using the credit-creation facilities of the IMF to produce the money to solve the climate crisis.

Keywords Loss-and-damage · Climate finance · COP negotiations · Paris Agreement · Green New Deal · IMF

© The Author(s), under exclusive license to Springer Nature Switzerland AG 2022

M. Scott Cato, *Sustainable Finance*,
https://doi.org/10.1007/978-3-030-91578-0_3

1 What Is 'Climate Finance'?

Climate finance has come to have a specific meaning in the context of the global climate negotiations conducted by the UN under the heading COP (Conference of the Parties—in this case 'parties' are countries that have signed up to the United Nations Framework Convention on Climate Change (UNFCCC). UNFCCC defines climate finance as 'local, national or transnational financing—drawn from public, private and alternative sources of financing—that seeks to support mitigation and adaptation actions that will address climate change'. It also notes that the two international climate treaties that have been agreed—the Kyoto Protocol (1997) and the Paris Agreement (2016)—have both called for those countries with greater financial power to provide financial resources to lower-income countries. And it notes that such finance should be directed towards both mitigation and adaptation. This definition provides a neat framework for this chapter where we consider how to fund the transitional investments needed to address the climate emergency, whether that financing should be public or private, and what it should best be invested in.

The UNFCCC is clear that while climate finance is finance 'that aims at reducing emissions and enhancing sinks of greenhouse gases [or adaptation to] negative climate change impacts', it may be public or private.[1] As we will see later, the question of who creates the finance and what conditions it has on it are crucial. The UNFCCC is clear that private finance should only be used for projects that benefit the community and economy but do not yield revenue streams for private investors. They add additional recommendations that private climate finance should:

- Partner with local companies and public sector organizations to transfer skills and share benefits.
- Build in long term training and education programmes for local people to learn the skills and build their own companies or organizations to do the work.
- Include women as workers, with decent work, and trainees.
- Provide the low-income countries with free access to knowledge and technology (global knowledge commons).
- Support the transition from global to local and regional trading.

[1] https://unfccc.int/topics/climate-finance/the-big-picture/introduction-to-climate-finance.

As we will see in Sect. 5, if these conditions are not met the addressing climate change may mean a subjugation of the countries of the Global South and the acquisition of power over their resources that feels like a new round of colonialism.

The UNFCCC process established a number of different financial mechanisms to provide the necessary climate finance. The first was the Global Environment Facility (GEF) established in 1994. At COP 16 (Paris, 2010) the Green Climate Fund (GCF) was set up, and made the main funding body of the UNFCCC. There are also two special funds: the Special Climate Change Fund (SCCF) and the Least Developed Countries Fund (LDCF), both managed by the GEF—and the Adaptation Fund (AF) established under the Kyoto Protocol in 2001.

Established following the 1992 Rio Earth Summit, the Global Environment Facility provides finance for a range of environmental crises, not just the climate crisis. Since 1994 it has made grants totalling more than $21.5bn and drawn in an additional $117bn in co-financing. Its finance is focused on the key UN environment conventions covering biodiversity, climate change, chemicals, and desertification and has 184 national governments in membership. By studying 4,574 projects implemented by the GEF they investigate how successful it was at leveraging additional finance. They find that the emerging economies are more successful at accessing additional finance than the lower-income countries and proposing that emerging countries should be required to secure more co-financing than lower-income countries.

As far back as the 2009 Copenhagen COP, the world's more economically powerful countries pledged $100bn in climate finance to the countries of the Global South. An academic analysis of this promise published in *Nature* finds that it was so ill-defined as to be impossible to hold countries to their commitments.[2] A promise made in haste to prevent the walkout of the countries more vulnerable to climate change has proved a poor basis to ensure they are compensated for their immense losses. They identify three main flaws in the definition of climate finance:

- Without a clear UN means of accounting, countries are free to decide what they consider to be climate finance—for example,

[2] Roberts, J.T., Weikmans, R., Robinson, Sa, et al. (2021), 'Rebooting a Failed Promise of Climate Finance', *Nature Climate Change*, 11, 180–182. https://doi.org/10.1038/s41558-021-00990-2.

Oxfam estimated that only a third of the public finance claimed by a 2020 OECD report, has actually been invested;

- The funds operate so as to disempower the very communities they were intended to support—funding channels are diverse and include developed countries' aid and export promotion agencies, private banks, equity funds and corporations, and lending and granting arms of multilateral institutions like the World Bank, none of which have oversight by the countries who need to be recompensed for loss and damage;
- Mitigation is prioritized over adaptation[3]—of climate finance flows that can be measured only about 20% has targeted adaptation, so rather than preparing countries for a climate-proofed future the money is mostly paying for repairing past damage.

2 The Loss-and-Damage Agenda

Much of the most controversial discussion at COP negotiations focuses around 'loss and damage' a phrase that, like climate finance itself, began with a generic sense but has since accrued a very specific sense.[4] As pointed out by the Grantham Institute[5]:

> The 'Loss and Damage' policy debate often focuses on the developing country context and this is reflected in the United Nations Framework Convention on Climate Change (UNFCCC) Decision 3/CP.18 preamble. The Loss and Damage debate has been contentious within the international climate negotiations because of questions of fairness and equity, and proving historical responsibility for climate change, in paying for the losses and damages associated with climate change. Developing countries have

[3] The mitigation vs adaptation debate is central to discussion on climate finance. In essence, mitigation means preventing the climate crisis from worsening, for example by investing in solar electricity; adaptation means investment to protect ourselves from the impacts of climate change that are already built in, for example through flood defences.

[4] Huq, Saleemul, Roberts, Erin, and Fenton, Adrian (2013), 'Loss and Damage', *Nature, Climate, Change*, 3, November.

[5] Grantham Institute (2021), 'What Is Climate Change "Loss and Damage"?' LSE Explainers. https://www.lse.ac.uk/granthaminstitute/explainers/what-is-climate-change-loss-and-damage/.

called for compensation from developed countries, while developed countries have sought instead to treat losses and damage as a sub-component of adaptation within the UNFCCC negotiations'.

Global climate negotiations are often stalled or broken by the struggle between the countries that have mainly caused climate change—by early industrialization powered by fossil fuels—and those countries that are suffering the earliest and most severe impacts. The loss-and-damage agenda acknowledges this and it was a breakthrough when, at COP19 in Warsaw in 2013, the Warsaw International Mechanism for Loss and Damage associated with Climate Change Impacts was established. This was followed by the establishment of the Santiago Network on Loss and Damage (SNLD) at COP25, an initiative that needs to be formalized at COP26 so that it can begin to mobilize the transfer of funds.

As long ago as 2017, Climate Action Network International (CAN) and Bond, a UK organization representing 450 UK NGOs working in international development, estimated that the cost of loss and damage is likely to exceed £1trn a year.[6] With climate impacts accelerating this is already probably an underestimate. It should be stressed that this funding for loss and damage is in addition to the $100bn pledge for climate-related financial transfers that is nowhere near being met. Some indications of the likely impact of the climate crisis are given in Fig. 1.

The agenda is politically challenging because its intention is to oblige the wealthy developed countries to take responsibility for their historic emissions and make recompense. Using the UK as an example, Harpreet Kaur Paul has calculated that the UK's historic emissions should require it to reduce its current emissions by 200% below 1990 levels by 2030. Since this is impossible, the UK should compensate the countries of the South by helping them invest to undergo a rapid sustainability transition.[7]

Saleemul Huq, director of the International Centre for Climate Change and Development (ICCCAD) in Bangladesh and lead author of the chapter on Adaptation and Sustainable Development in the third assessment report of the Intergovernmental Panel on Climate Change,

[6] Joint Submission on the Strategic Workstream on Loss and Damage Action and Support, to UNFCCC (2017): Microsoft Word—CAN Bond Joint Submission on the Strategic Workstream on Loss and Damage Finance.docx (unfccc.int).

[7] Kaur Paul, H. (2021), *Towards Reparative Climate Justice: From Crises to Liberations* (CommonWealth).

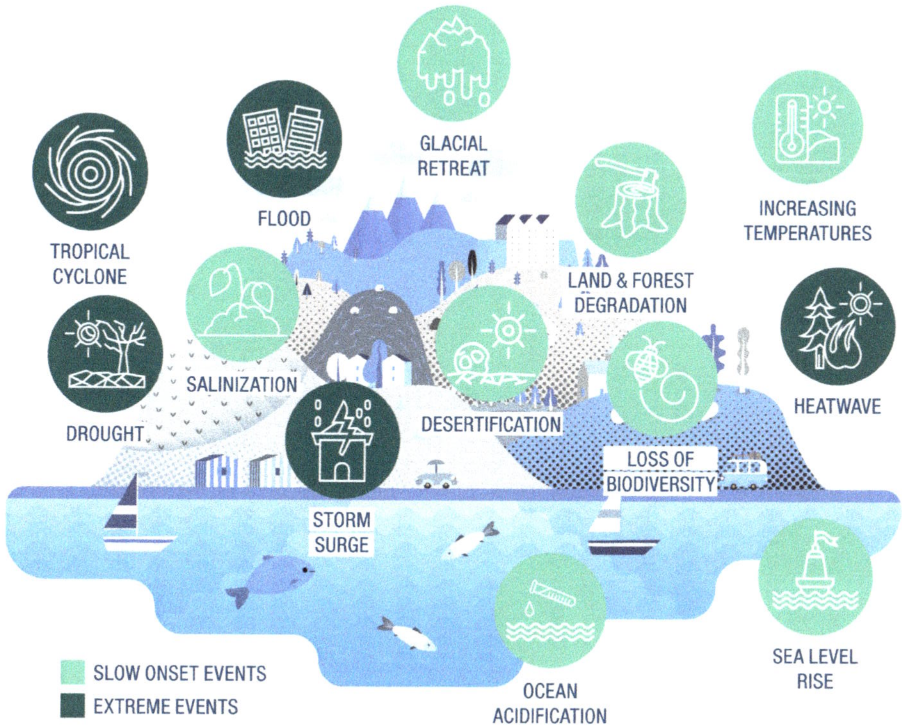

Fig. 1 What does loss and damage look like (*Source* Author's graphic redrawn by Angela Mak)

has stressed the centrality of the loss-and-damage agenda in deciding the success or otherwise of global climate negotiations. He has written that 'Dealing with loss and damage from human induced climate change in the upcoming COP26 in Glasgow, Scotland in November 2021, will be a make-or-break issue for the most vulnerable developing countries'.[8]

The key player here is the US which, since the election of President Biden and the designation of global diplomat John Kerry as Climate Envoy, is now playing a leadership role in the global climate negotiations. Under President Trump the US was absent if not hostile to climate action, and historically it has always acted as a block on serious action on funding for loss and damage.

As far back as Copenhagen (COP15, 2009), the issue of compensation for those suffering climate impacts caused by past actions of the big polluters was put on the agenda of the UNFCCC, causing US negotiator Todd Stern to reject the idea: 'We absolutely recognize our historic role in putting emissions in the atmosphere up there that are there now … but the sense of guilt or culpability or reparations, I just categorically reject that'. In fact, the US operated a reverse policy of withdrawing aid from countries that were 'difficult' over climate negotiations.[9]

While much of the focus on discussion about reparations for climate damage focuses on cash payments and investments, there is also the possibility of technology transfer, more like a compensation in kind. Given that the globally powerful economies developed their technological advantages through the burning of fossil fuels, it seems not only pragmatic but also reasonable that they should share the knowledge gained in this way with the countries that are bearing the brunt of the climate crisis they caused.

A successful resolution of the loss-and-damage agenda looks likely to be a make-or-break issue at COP26 and beyond. At the time of writing, how to fund repairs is not even on the negotiating agenda, something that will be challenged by Bangladeshi Prime Minister Sheikh Hasina, as chair of the Climate Vulnerable Forum. The Climate Technology Centre Network is a step in this direction, providing technical assistance to developing countries to tackle climate change. Run by the

[8] 'Dealing with Loss and Damage in COP26', *The Daily Star*, 17 August 2021. https://www.thedailystar.net/opinion/politics-climate-change/news/dealing-loss-and-damage-cop26-2041965.

[9] Sealey-Huggins, Leon (2017), '1.5 oC to Stay Alive': Climate Change, Imperialism and Justice for the Caribbean', *Third World Quarterly*, 38:11, 2444–2463.

UN Environment Programme rather than UNFCCC directly, it has bases in many developing countries that can request technical assistance to support infrastructure essential to the climate transition, such as solar energy systems, rainwater harvesting and so on.

3 Green New Deals

Although governments have been providing incentives for the green economy in fairly limited ways for decades now, the Covid crisis appears to have caused a burgeoning of new public investment in the sustainable economy, especially in the US and EU. Although government investments in the sustainability transition are not formally defined as 'climate finance' it seems important to cover them here, particularly because the way they are funded indicates the injustice of a world where some countries can find a trillion dollars down the back of the sofa but others need to lobby for decades to repair climate damage caused by those same countries.

The idea of a Green New Deal has been around for more than a decade, since a group of green economists and policy-makers launched it in London on the back of Colin Hines's work with the New Economics Foundation.[10] From an economic perspective it proposed a triple win: investment in the green transition, jobs for the workless, and stimulus for the economy. It was almost totally ignored, with the government instead proposing its Green Investment Bank, an ill-conceived and short-lived proposal (discussed further in Chapter 6).

But ten years on the call has been heard around the world, most resoundingly in the US, perhaps because it explicitly echoes the largest Keynesian response to the Depression: Roosevelt's New Deal programme of infrastructure investment and job creation. The coincidence of global evidence of the reality of the climate crisis with an unprecedented economic contraction caused by Covid-19 gave rise to calls across the world for a green stimulus to support the world in 'building back better'. It would be churlish not to acknowledge the shift in consciousness and the significant policy shift this has led to, but it is important also to cast a critical eye over the reality of these green stimulus promises.

Most impressive in mere size is President Biden's Build Back Better Plan that is projected to amount to an extraordinary $4.5trn over the

[10] Original Green New Deal Report (2008). http://www.neweconomics.org/sites/neweconomics.org/files/A_Green_New_Deal_1.pdf.

next decade,[11] a significant portion of it directed towards green stimulus in the form of investments in green infrastructure and especially to decarbonize the US energy system, retrofit homes, and improve energy efficiency of buildings. At the time of writing only the first tranche of this Covid recovery plan—worth $1.9trn—has been passed into law and the need to gain political agreement has caused a shift in emphasis from environment to health and social programmes.

Under the rubric 'Repair and prepare for the next generation' EU Commission President Ursula von der Leyden announced the European Union's Covid recovery plan in May 2020. This has since been reframed under the overarching policy objective of the EU: the European Green Deal. The Green Deal is ambitious, promising at least €1 trn (£852bn) of green stimulus of which €503bn, should come directly from the EU budget. The remainder comes via the InvestEU programme, a combination of public and private funding with national match-funding. This will underpin riskier lending by the EIB, now rebranded as Europe's 'climate bank'.[12]

Opposition within the EU comes, as usual, from the eastern European countries, who are still defending their energy-intensive industries and especially, in the case of Poland, their coal industry. The Just Transition fund is the key policy to persuade these countries to take climate action seriously and has real teeth, since countries will only receive half of their share if they refuse to sign up to the NZC by 2050 target, a weaker target than the 100% exclusion the Commission proposed, but still a powerful incentive to change. Achieving this target will require all EU member to states to invest rapidly in energy-efficient buildings, decarbonization of electricity generation; and rolling out cleaner, cheaper, and healthier forms of private and public transport.

So coming to the question we hear so frequently from climate delayers: what about China? China's government has promised 3.6trn

[11] Zandi, Mark (2021), 'Biden's Build Back Better Plan Will Improve Nearly Every Community in America', *CNN Business*.

[12] EU Commission (2020), The European Green Deal Investment Plan and Just Transition Mechanism explained: The European Green Deal Investment Plan and JTM explained (europa.eu).

Yuan ($500bn) in Covid stimulus, focused on supporting infrastructure, especially 5G and electricity transmission.[13] It is unclear yet how this will balance general infrastructure development against, for example, charging points for electric vehicles, but the absence of an annual growth target is being met with cautious optimism. Initial analysis by Carbon Brief suggests that China's high-energy sectors, including coal power, and cement, are bouncing back faster than other sectors.

Space limitations mean I have only been able to focus on the largest players here, but a full account of a wide range of countries and their green stimulus packages has been undertaken by Carbon Brief.[14] To answer the even more important question—how green are these stimulus packages in reality rather than in rhetoric—we turn to an analysis by Vivid Economics. Their rather depressing conclusion is that 'the $17.2tr of COVID-19 recovery money injected into the global economy to date continues to have a net negative impact on climate and nature', according to their 'Greenness of Stimulus Index' (GSI) report.[15]

They analysed the economies of the G20 countries plus another 10 countries and found that $1.8trn worth of investments will have a net positive impact (defined as reducing climate emissions or enhancing biodiversity in areas relating to the environment, energy, transport, industry, agriculture, and waste) while $4.8trn will damage the environment. Ten countries had improved their score since February that year with Denmark out in front and US, Norway, and South Africa improving their performance while Russia, Turkey, and Singapore scored worst. The analysts found that nature and biodiversity had been particularly neglected.

Both China and India are improving their performance over time because of their focus on reducing CO_2 emissions, but both are also investing in coal. The authors regret that, in spite of the hot air, Biden's

[13] State Council of the People's Republic of China (2020), 'Work Report Delivered by Premie Li Keqiang', Third Session of the 13th National People's Congress, 22 May: Full Text: Report on the Work of the Government (www.gov.cn).

[14] Carbon Brief (2020), *Coronavirus: Tracking How the World's 'Green Recovery' Plans Aim to Cut Emissions*. https://www.carbonbrief.org/coronavirus-tracking-how-the-worlds-green-recovery-plans-aim-to-cut-emissions.

[15] Beyer, Jeffrey and Vandermosten, Alice (2021), *Greenness of Stimulus Index: An Assessment of COVID-19 Stimulus by G20 Countries and Other Major Economies in Relation to Climate Action and Biodiversity Goals* (Vivid Economics). https://www.vivideconomics.com/wp-content/uploads/2021/07/Green-Stimulus-Index-6th-Edition_final-report.pdf.

$1.9 trillion American Rescue Plan fails to specifically target climate change and biodiversity issues, but does include investments in public transport and improvements to water, sewage, and energy efficiency. They conclude that:

> Overall, only Canada and parts of Europe oriented their stimulus in a way that significantly shifted their trajectory, thanks to a concerted effort from early on in the crisis. Denmark and Canada made the largest overall efforts to reorient their economies through the stimulus spending, with the European Commission spending and national-level stimulus packages in the UK, France, Germany, Finland, Spain and Sweden achieving strongly positive outcomes. Other more advanced economies – such as Japan, South Korea, Italy and Australia – made some efforts but did not manage to achieve a transformational shift through their stimulus.

It is noticeable that the US has invested vast sums in Covid recovery. There is a simple reason for this: the dollar is the world's currency so can be created almost at will and at virtually no cost to US taxpayers. Other countries whose currencies are held as reserves—especially the Japanese Yen and the pound sterling—also have much greater freedom to create money in times of crises. The EU is more constrained because—although the euro is a powerful currency—there is disagreement between the member states on monetary policy, with Germany in particular being reluctant to use the power of direct money creation. The crucial point here is that countries without a powerful currency—most of the countries in the world—are forced to borrow to make these sorts of investments. This is a major source of inequity that I discuss further in Chapter 6.

4 Climate Reparations or Empire 2.0

The climate change that is with us now is the result of CO_2 emissions from past decades when the countries of the West began their industrial revolutions. So to achieve climate justice we have to ensure that those of us whose prosperity was created through these polluting processes now come forward to ensure those who are suffering from the climate crisis are protected. This is in return for our heavy responsibility for causing the climate crisis. As Harpreet Kaur Paul writes, we need a 'fair shares' approach to climate change: 'a fair share of responsibility for countries like the UK - early industrialisers with historic responsibility for our current crisis - necessitates both domestic and international action. Within and between countries, this action must acknowledge that the wealthy have the highest degrees of resilience to climate change shocks, but also the greatest responsibility for emissions'.[16]

But more fundamentally, the countries that carry the largest burden of historic CO_2 emissions are the same countries that grew rich through colonial exploitation: there is an inextricable link between colonialism and historic emissions. The Civil Society Review of Loss and Damage is clear that[17]:

> Colonialism and the fossil fuel era reconfigured the world economy. The Indian subcontinent's share of the global economy shrank from 27 to 3 per cent between 1700 and 1950 and it is estimated that the UK extracted approximately USD$45 trillion from its colonial rule of the Indian subcontinent alone. China's share shrank from 35 to 7 per cent. At the same time, Europe's share of the global economy exploded from 20 to 60 per cent.

Countries that are bearing the brunt of the increasing power of hurricanes were themselves made vulnerable by economies that were formed in the interests of enslavers. Haiti is a prime example: 'the poorest country in the Western hemisphere... it was forced to pay reparations to France for having the temerity to throw off its colonial master and establish itself as the first Black republic of the "new world". Other Caribbean societies continued to be manipulated by colonial powers and 'forced to compete on a deeply uneven playing field when they gained their "independence"'.

The exploitation of the natural world and our current climate emergency are directly connected to the development of an extractivist, capitalist economy that was forged through the systemic enslavement of African peoples and the parallel system of colonial occupation, warfare, land-grabbing, oppression, and resource extraction. This was a system that not only separated people from their ancestral rights to land and its stewardship, but also portrayed that land and its people as a resource

[16] Harpreet Kaur Paul (2021), *Towards Reparative Climate Justice: From Crises to Liberations* (London: Common-wealth). 6071e27f9e138da86620f637_CW_GND-Reparations-Harpreet.pdf (webflow.com).

[17] 'Can Climate Change Fuelled Loss and Damage Ever be Fair' (2019), Report signed by a global group of civil rights and environmental organisations, p. 11: Can Climate Change Fuelled Loss and Damage Ever Be Fair? (civilsocietyreview.org).

to be exploited for the accumulation of wealth that profited the monarchies and nations of the Global North. So climate-related reparations for colonialism and the trafficking of enslaved people is not about charity but justice.

And if we do not arrange climate finance to repair loss and damage appropriately then we will merely be repeating colonial exploitation in a new guise, in a financialized form. For example, using the land of countries of the Global South to absorb our emissions through some 'offsetting' or 'carbon credits' scheme would be a new form of enclosure and exploitation by the wealthy countries that were their former colonizers. Any use of the land of countries of the South—that have preserved their environments much more successfully than the early industrializers—would constitute a form of neocolonialism.

Esther Stanford-Xosei and Nicki Frith are co-founders of the International Network of Scholars and Activists for Afrikan Reparations (INOSAAR). They have developed the concept of 'planet repairs' to deepen and extend the discussion of reparations in an environmental context. Their definition resulted from a dialogue between them that is firmly grounded in the thinking of the Pan-Afrikan Reparations Coalition in Europe of which Esther is a founding and leading member. The following definition they offer is firmly grounded in that organization's thinking[18]:

> Planet Repairs refers to the need to proceed from a standpoint of pluriversality that highlights the nexus of reparatory, environmental and cognitive justice in articulating the need to repair holistically our relationship with, and inseparability from, the earth, environment and the pluriverse. It means giving due recognition to Indigenous knowledges in contrast with western-centric Enlightenment ideals that separated humanity from nature and devalorized Indigenous systems of knowledge in order to justify exploitation for capital accumulation.

They connect the alienation between North and South with the alienation of humanity from nature and trace the origins of both to the era of African enslavement and colonialism.

[18] Frith, Nicola and Stanford-Xosei, Esther (2022, forthcoming), 'Reparations Activism in the UK: A Pan-African Journey Towards Planet Repairs', in *Slavery, Colonialism, and Reparations*, ed. by Adekeye Adebajo and Anita Montoute (University of Johannesburg: Institute for Pan-African Thought and Conversation).

The purpose of the campaign for planet repairs is that climate justice should be placed within a historical context of centuries of exploitation of countries of majority world by the developed countries. This exploitation is linked to their greater responsibility for historic emissions that are causing loss and damage across the Global South. This framing of the responsibility wealthy countries carry for the historic emissions in terms of reparations and in the language of 'planet repairs' could be a constructive way forward to achieving an agreement in Glasgow. But it would require the economically powerful countries—largely those of the G7—to accept their historic responsibilities and finance the costs of loss and damage.

5 We Can Afford to Save the Climate

Absurdly, we are left in a situation where the question 'Can we afford to save the planet?' seems like a meaningful one. There is no need for this. Money is a creation of the human imagination and, as a global community, we have our own bank that can create the money we need to save life on earth. In Chapter 6, I explain in a bit more detail how the IMF, while not being the world's bank in the way any of us would design to maximize equity or sustainability, nonetheless does have the power to create the finance needed to resolve the issue of loss and damage, as called for by the IIED (International Institute for Environment and Development).

Durand and colleagues explore some ways in which the finance for loss and damage could be found.[19] A number of their proposals include funding the costs through forms of insurance or bonds. This strikes me as rather unhelpful, since the inevitability of the damage means that the risk-sharing design of insurance misses the point, and it is hard to imagine how damage could generate an income stream to repay purchases of a bond. However, they do have some proposals that might be feasible. A Financial Transactions Tax has become something of an old chestnut in political circles—a small charge on the vast flows of cash in the global financial casino—and has been spent many times over long before it has

[19] Durand, Alexis, Hoffmeister, Victoria, Timmons Roberts, J., Gewirtzman, Jonathan, Natson, Sujay, Weikmans, Romain, and Huq, Saleemul (2016), 'Financing Options for Loss and Damage: A Review and Roadmap Climate and Development Lab (CDL)', Brown University International Centre for Climate Change and Development (ICCCAD). bro wncdl-icccadfinancinglossanddamagepaperdraft.pdf (unfccc.int).

been politically agreed, but is certainly worth considering. An international passenger levy was proposed to the UNFCCC in 2008 by Maldives on behalf of the 48 country LDC group of nations and has the benefit of immediate moral appeal, while a carbon tax (as discussed in Chapter 4) is a powerful tool but gaining political agreement for the yield to be shared globally would be more difficult.

Under the heading 'other tools' they reach a proposal that seems to be the most promising: the issue of SDRs as the basis of an allocation directed towards the countries suffering loss and damage. They note that this was proposed by George Soros in 2009 and supported by then IMF Managing Director Dominique Strauss-Kahn before 'disappearing from the discourse'. However, the proposal has recently been revived by development charity CAFOD and by a group of European charities under the auspices of the Robin Hood Tax campaign.[20] As discussed in Chapter 6, SDRs represent a way of creating global liquidity, resolving the problem identified above—that countries without reserve currencies cannot gain the benefit of QE to fund their climate policies. CAFOD drew attention to the IMF's creation of a record amount of these reserve assets in August 2021 and suggested the UK should use its share to fund transfers related to loss and damage.

The International Institute for Environment and Development (IIED) went further and, at a conference held in July 2021, suggested that the new issue of SDRs should be used to fund climate action.[21] The potential scale of the spending power that could be unleashed is striking: the $650 billion SDRs that were issued for green recovery are equivalent to 65 times the size of the current Green Climate Fund (GCF). If these were directed towards the countries suffering the worst of climate impacts they could provide the funding for loss and damage. In a blog with the former

[20] CAFOD (2021), *Using the United Kingdom's SDRs to Tackle Covid-19 and Climate Change, Discussion Paper* (London: CAFOD): Using the UK SDRs. CAFOD discussion paper May 2021.pdf; Robin Hood Tax campaign (nd), 'Unpacking Finance for Loss and Damage: Lessons from COVID-19 for Addressing Loss and Damage in Vulnerable Developing Countries'. https://www.robinhoodtax.org.uk/sites/default/files/Unpacking%20Finance%20for%20Loss%20and%20Damage%20brief%201%20-%20Lessons%20from%20Covid.FINAL_0.pdf.

[21] IIED (2021), 'A Green Recovery for Inclusion: Debt Relief and Special Drawing Rights for climate action', climate held on 7 July: A green recovery for inclusion: debt relief and special drawing rights for climate action | International Institute for Environment and Development (iied.org).

Finance Minister for Pakistan, Shamshad Akhtar, Paul Steele explains how this might be possible[22]:

> Some in the IMF would be willing to support such re-allocated 'SDRs for green recovery' – for example the more climate-vulnerable a country the larger their reallocation – or by linking SDR reallocations to spending on pro-poor and growth-enhancing climate resilience or biodiversity investments and policies. These $650 billion SDRs for green recovery would represent 65 times the size of the current Green Climate Fund (GCF) and if judiciously distributed could finance the much-needed climate adaptation and mitigations needs of low-income countries.

It is problematic and unjust that, every time the world's bank issues new currency, it is allocated according to the shareholdings in the bank, meaning that the richest countries receive the largest allocations. However, as Ellmers and colleagues point out, the countries that receive the largest allocations are quite at liberty to reallocate them to the countries suffering most from climate impacts.[23] The key to funding loss and damage is persuading these countries that they should forego that free lunch and distribute it to the countries suffering loss and damage in recompense for the damage caused by their historic emissions. This does feel too good to be true, but with a climate-concerned President in the White House, it could be a way for countries to finance loss and damage without having to negotiate with their national finance ministers about what needs to be cut to make this possible.

6 Principles of Climate Finance

As we have seen, addressing the climate crisis comes with enormous potential risks that some of those most vulnerable to impacts will be disempowered—even exploited—by policies framed as solving their problems. To avoid this the IIED has provided some useful principles to guide

[22] Steele, Paul and Akhtar, Shamshad (2021), 'Here's How to Propel a Green Recovery for the Poorest', Opinion, Thomson Reuters Foundation: Here's how to propel a green recovery for the poorest (trust.org).

[23] Ellmers, B. (2020). 'Financing Sustainable Development in the Era of COVID-19 and Beyond: An Analysis and Assessment of Innovative Policy Solutions', Bonn: Brot für die Welt, Global Policy and MISEREOR [online] Available at: https://www.globalpolicy.org/images/pdfs/Briefing_1220_FSD_Covid-1.pdf.

Table 1 Principles of Good Climate Finance

Principle	*Explanation*	*Example*
Subsidiarity	Making decisions as close as possible to those most affected to find solutions responsive to local conditions and avoiding excluding the marginalized	The Climate Investment Fund's Pilot Programme for Climate Resilience (PPCR) prioritizes a community-driven resilience approach in Zambia and has facilitated strong local stakeholder engagement in Tajikistan
Robust decision making	This requires involvement and training of local people and respecting their local and traditional knowledge	GCF project in Bhutan combines traditional and indigenous knowledge with local weather, seasonal and climate information
Patient and predictable	Timescales long enough to enable building capacity and to permit risk-taking	Antigua and Barbuda's innovative EDA project seeks to build the capacity of at least three 'whole-of-society' on-financing mechanisms
Flexibility	Because outcomes and impacts are unpredictable	Limited success so far
Risk-taking	Because experience is limited	The DGM (Dedicated Grant Mechanism) uses an 'empowerment pathway' approach, prioritizing local organizations' fund-management skills, community representation and ability to raise local issues at national and global levels, and embracing the failure of subproject objectives as learning-by-doing
Converged	While coordination is now happening, the wide variety of climate-related vulnerabilities requires converging actions by different funders	No examples as yet

the investment of climate finance, reproduced in Table 1.[24] It is clear that climate solutions need to respond to very local situations and to empower communities and their knowledge of local places and systems.

[24] Patel, Sejal, Soanes, Marek, Rahman, Feisal, Smith, Barry, Steinbach, Dave, and Barrett, Sam (2020), 'Good Climate Finance Guide: Lessons for Strengthening Devolved Climate Finance', IIED Working Paper. https://pubs.iied.org/sites/default/files/pdfs/2021-01/10207IIED.pdf.

And that the unpredictability of climate impacts requires a particular form of consideration that may be innovative for some funders.

It is useful to have these guiding principles in terms of implementation, but we also need to ask whether it can ever be appropriate to make loans, rather than grants, for climate investment. Can we imagine any situation where a loan would not simply extract value from already poor countries and exploit their people even further?

Oxfam provided an analysis of climate finance transfers in 2020 and found that public climate finance, i.e. grants from governments rather than loans, increased from $44.5bn per year in 2015–16 to an estimated $59.5bn per year in 2017–2018. However, deceptive reporting means that this figure is an overestimate once repayments, interest, and other factors are taken into account. The deceptive nature of claims about climate finance is illustrated in Fig. 2.

Taking account of these factors, Oxfam estimates that[25]:

- Specific net assistance is much lower than reported figures, increasing slightly from $15–19.5bn per year in 2015–16, to $19–22.5bn per year in 2017–18.5
- Provision of concessional loans and other non-grant instruments is estimated to have increased from $18.5bn per year in 2015–16 to $22bn per year in 2017–18;
- An estimated 40% of public climate finance overall is non-concessional meaning the terms are not generous enough to qualify as Official Development Assistance (ODA);
- This proportion had increased from 30% in 2015–16;
- Over 40% of what is reported as 'public finance' is actually in the form of loans from multilateral development banks (MDBs).

So while the richer countries will claim that they are on track for the $100bn target, much of this is based on loans that will only push the countries of the majority world deeper into debt. Oxfam considers the provision of climate finance via loans and other financial instrument that impose obligations on recipients is 'an overlooked scandal'. Their moral conclusion could not be clearer: 'the world's poorest countries and

[25] Oxfam (2020), *Climate Finance Shadow Report 2020: Assessing Progress Towards the $110billion Commitment* (Oxford: Oxfam): Climate Finance Shadow Report 2020: Assessing progress towards the $100 billion commitment (openrepository.com).

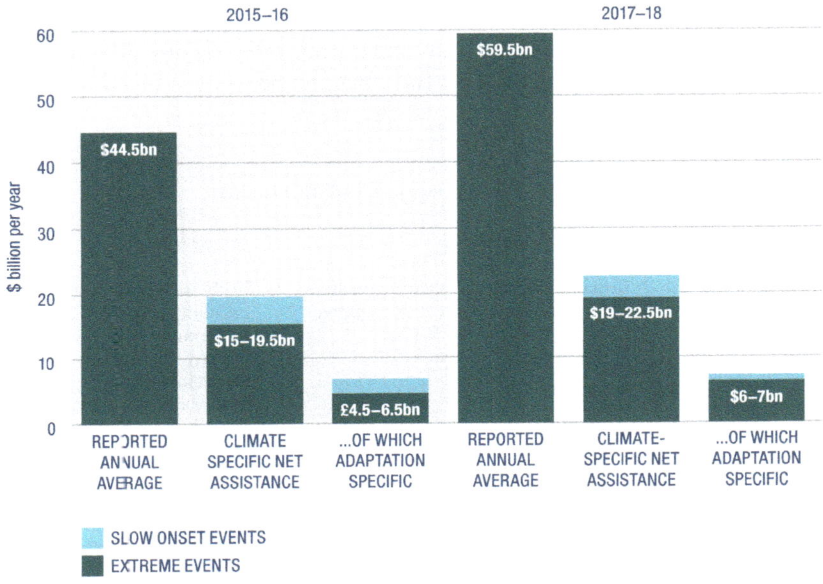

Fig. 2 The real value of climate finance (*Note* Developed countries' reported climate finance versus Oxfam's estimate of 'climate-specific net assistance' [2017–2018 and 2015–2016 annual averages], taken from the Oxfam Shadow Climate Finance Report 2020. *Source* 2017–2018 numbers—Fourth Biennial Reports [2020] and OECD (2020a). See Box 1 for details of how climate-specific net assistance is calculated. Note 21 sets out how total reported public climate finance was estimated for 2017–2018. 2015–2016 numbers—reported climate finance as set out in OECD [2019a], and see T. Carty and A. le Comte [2018] for climate-specific net assistance estimates, which have been adjusted in line with reported climate finance estimated in OECD [2019a]. Author's graphic redrawn by Angela Mak)

communities should not be forced to take out loans to protect themselves from the excess carbon emissions of rich countries. Finance that should be helping countries respond to climate change should not be harming them by contributing to rising – and in many countries, unsustainable – debt levels'.

So insisting on public vs private finance and grants rather than loans is an essential condition for achieving climate justice. But it is not sufficient: we need a series of further principles to mitigate the political risks from climate finance. Any successful resolution of the loss-and-damage issue requires a much tighter and more politically secure definition of climate finance. Some suggestions for what the principles might look like are outlined in Table 2.

Table 2 Political Principles for Equitable Climate Finance

Principle	*Explanation*	*Example*
No displacement to other sectors	Climate finance investments should not cause negative spillover effects on environments	With such a strong focus on climate it is possible for environments to be destroyed in other ways, for example the biodiversity lost when a habitat is planted with trees as part of a carbon credit scheme
No undermining of local livelihoods	Climate finance should not mean the use of resources—perhaps outside the market—that are essential people's livelihoods	A wealthy company might buy land for a tree-planting scheme to offset its emissions when that land previously provided grazing for local people's cattle
No increase in financial inequality within or between countries	Finance invested in any country should not result in resources or economic value being extracted from that country	Finance needs to flow from the wealthy nations to the poorer, not the other way around; in practice this means climate finance must be based on grants, not loans
No increase in political inequality within or between countries	Climate finance invested in any country should be controlled by community actors in those areas where the investments are made and who have suffered from climate-related loss and damage	As discussed in Table 1, domestic or foreign governments must be prevented from acquiring funds that they use in a way that reduces the resources of local communities, for example displacing them from their land
No increase in social inequality within or between countries	These community actors should truly represent the local population, especially in terms of gender	Self-explanatory: climate finance could change social power relations and this must not be allowed to reverse advances in human rights, especially by women

What would this mean in terms of outcome from the COP negotiations that would satisfy the requirement for planet repairs? We would suggest the following as a basic minimum:

- Immediate debt relief—for indebted countries who face the current climate emergency;
- Loss and Damage listed as a permanent COP agenda item so that the emphasis of climate finance is shifted towards repairing the damage caused by nations to their contributions to repair that damage;
- A new global public finance goal specifically for adaptation—as a component of the new collective finance goal starting in 2025, when the $100bn commitment will be succeeded.
- A financing facility to deliver public climate financing and new and innovative sources of financing to address the loss-and-damage agenda;
- A commitment to urgently increase grant-based public climate finance, in particular to LDCs and SIDS.
- A commitment to a Loss and Damage Gap Report with regular updates and clear targets to close the gap.

CHAPTER 4

Sustainable Finance: The Policy Framework

Abstract Rather than tackling climate change, most governments are still subsidizing fossil fuels. The EU has taken a global lead on developing a legislative framework for sustainable finance that is introduced here, while developments in Japan and the US as also covered. The next section considers policy responses to the urgent need to put a price on carbon and make polluters pay. There is widespread agreement that addressing climate change means putting a rising price on carbon but no agreement on how best to achieve this, whether by agreeing on a carbon tax on a global basis—or within smaller communities of nations—through a carbon trading scheme. I explore a proposal to use the regulatory and legislative powers of policy-makers to eliminate unsustainable assets from the global economy. Finally, we explore how the financial system needs to change to address the biodiversity crisis.

Keywords Fossil fuel subsidies · Carbon trading EU ETS · Carbon tax · Biodiversity crisis · Das Gupta Review

© The Author(s), under exclusive license to Springer Nature Switzerland AG 2022

M. Scott Cato, *Sustainable Finance*,
https://doi.org/10.1007/978-3-030-91578-0_4

1 Feeding the Beast

Campaign-based action against fossil fuel investments has grown naturally out of shareholder activism and has been energized by the disturbing truth that most governments are subsiding the fossil fuel industry with one hand while they are signing pledges to reach net-zero with the other.

While this seems incongruous, inconsistent, immoral even, in a global economy that has been literally powered by fossil fuels for the past century, it is not surprising that this historic link between government industrial policy and the fossil fuel industry is strong. The fossil fuel industry may be a dinosaur whose lifespan is limited but it is still a powerful dinosaur that spends vast amounts of money on lobbying and funding politicians across the world. As Fig. 1 shows, where countries have fossil fuel assets their governments have given them tax breaks or grants to encourage their growth, and in most cases continue to do so. Figures from the International Energy Agency (IEA) show that the three countries that provide the largest value of subsidy/GDP are Iran, China, and India, followed by Saudi Arabia, Russia, and Algeria. And in spite of its pride in its Green Deal, the EU is still providing subsidies to the fossil industry with nearly a third of the €159bn spent on energy subsidies going to fossil fuels.

IEA data show that such subsidies fell in 2020, although more as a result of low demand for oil caused by the coronavirus pandemic than because of concern about the climate crisis. Subsidies come in the form of direct grants to the industry for new exploration or giving them preferential tax treatment or through government intervention to keep fuel prices lower for consumers. The IMF is clear that such subsidies ultimately benefit the wealthy, who use more energy, but the link between standard of living and pricing carbon appropriately is crucial when developing effective policies to reduce CO2 emission.

Rempel and Gupta[1] regret the fact that the much vaunted Paris Agreement neither referenced nor targeted fossil fuels as a key cause of the climate crisis 'because fossil fuels are a politically sensitive "hot potato" that even investors, stock brokers and academics hesitate to publicly

[1] Rempel, A. and Gupta, J. (2020), 'Conflicting Commitments? Examining Pension Funds, Fossil Fuel Assets and Climate Policy in the Organisation for Economic Co-operation and Development (OECD)', *Energy Research and Social Science*, 69, 101736. https://doi.org/10.1016/j.erss.2020.101736.

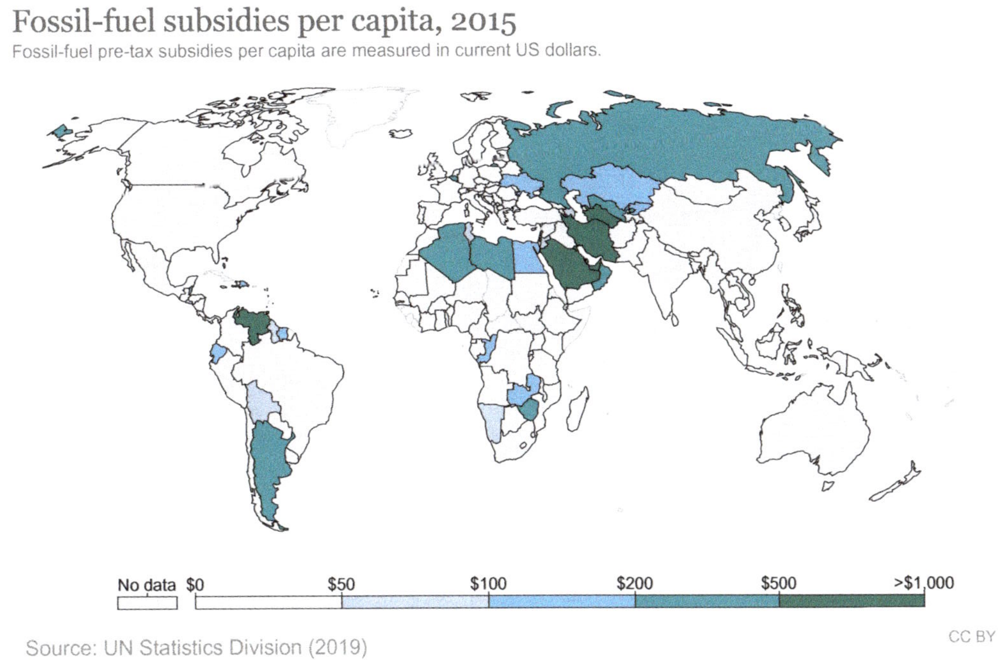

Fig. 1 Fossil fuel subsidies per capita, 2015 (in $US) (*Note* Fossil fuel pre-tax subsidies per capita are measured in current US dollars. *Source* Based on data from Our World in Data: All Our World in Data is completely open access and all work is licensed under the Creative Commons BY license. We all have the permission to use, distribute, and reproduce in any medium, provided the source and authors are credited; thanks to Wikimedia Commons for making this graphic available free of charge [A/w available open source online here: https://commons.wikimedia.org/wiki/File:Fossil-fuel_subsidies_per_capita,_OWID.svg])

discuss'. I can certainly vouch for this from my time as an MEP, when any attempt to even label fossil fuels as unsustainable was blocked by eastern European EU members who rely on these fuel sources not only for electricity production but also to fund their public pension funds. This continuing reliance on fossil fuels is imperiling any possibility of staying within a 2 °C warming trajectory since we are on course globally to 'produce about 50% more fossil fuels by 2030 than would be consistent with a 2 °C pathway and 120% more than would be consistent with a 1.5 °C pathway'.[2] In the run-up to the COP26 climate talks at the end of 2021, the rhetoric has built on the need to abandon coal on a global basis for electricity generation but this is somewhat hypocritical coming from countries that still subsidize their own fossil fuel industries.

2 Making the Polluters Pay

Since we live in a market economy the most powerful tool to shift the economy away from, first, fossil fuels, and then ecologically damaging activity more generally is to increase the price of undertaking those activities. In the case of climate change this is what has long been referred to as 'putting a price on carbon'; more widely it means using the tax system to ensure that the most environmentally supportive action for a business to take is also the one that makes sense in terms of their bottom line. The Stern Review was clear that 'Creating a transparent and comparable carbon price signal around the world is an urgent challenge for international collective action' (p. 530).

Neoclassical economists will sometimes refer to this as 'internalising and externality'. In theoretical economics, the problem of pollution arises because the cost of producing it is not included in the cost calculations of the firm so they have no compunction about producing as much of it as they can. As a green economist, I struggle with the idea that there is anywhere 'external' to our beautiful planet where pollution can be put safely out of sight and mind. But I think we can all agree that if polluters had to pay a high and rising price for their noxious emissions they would rapidly reduce or eliminate them.

[2] SEI, IISD, ODI, Climate Analytics, CICERO, and UNEP. (2019). 'The Production Gap: The Discrepancy Between Countries' Planned Fossil Fuel Production and Global Production Levels Consistent with Limiting Warming to 1.5 °C or 2 °C', cited by Rempel and Gupta (2020).

Creating a carbon price is a way of implementing the polluter-pays principle in the area of climate change: companies will no longer be able to treat the global atmosphere as a free dumping ground. The first debate is about where the policy is implemented—this is the so-called upstream vs. downstream debate. Upstream we have the producers, so we might impose a tax on them, for example, when they extract fossil fuels from the ground. At the other end of the chain—downstream—we have consumers, whose emissions might be controlled through giving them a limited allowance per year, for example. Upstream solutions tend to be cheaper, since there are fewer producers, but how can we be sure that the costs will all be passed on to consumers? On the other hand, downstream solutions, involving millions of consumers, are expensive to administer but place the responsibility on citizens to change their individual behaviour.

There have essentially been two methods for pricing carbon: creating a market for pollution and requiring those who produce it to purchase permits to produce it, permits which are then made tradable. We will consider the effectiveness of such a solution here before moving on to consider the carbon tax that is the second proposal.

Market-Based Pollution Permit Trading

A permit trading system to control pollution is designed to impose a limit on the amount of the pollution that can be emitted and then permit those who produce emissions to trade between themselves the right to do so; this explains a term that is sometimes used to describe such schemes: cap and trade. Those who support a system of emissions trading argue that it is efficient, since it ensures that those who make the reductions will be those who can do so most cheaply, and they will then sell their emissions rights to others, for whom that is a cheaper solution than reducing their own emissions. Such a scheme would (its supporters argue) also have the advantage that it would follow naturally from fixed caps negotiated internationally, and would provide a simple mechanism for governments to implement these caps nationally within fixed aggregate limits.

Although the US federal government introduced a trading scheme to reduce sulphur dioxide emissions following the passage of the Clean Air Act in 1990, such schemes are usually discussed in connection with reductions in climate-damaging CO_2 emissions. And the largest and most established such scheme is the EU Emissions Trading System or ETS.

The ETS is based on the concept of 'cap and trade', meaning that for companies within the system the total amount of CO_2 that can be emitted is capped, with the cap reducing over time. Companies within the system have to surrender emissions permits to match the CO_2 they produce. The EU ETS covers a wide range of industrial sectors including CO_2 emissions from electricity generation, energy-intensive industries including oil refineries, steel-works, and production of iron, aluminium, metals, cement, lime, glass, ceramics, pulp, paper, cardboard, acids, and bulk organic chemicals. It also covers the emission of other GHGs (greenhouse gases) including nitrous oxide and prefluorocarbons. There are exemptions for smaller companies and the whole aviation sector was exempted until 31 December 2023, except for flights within the European Economic Area.[3]

Critics have argued that the scheme is flawed because heavily polluting industries were given free allowances after intense lobbying while other sectors—including aviation—were left out of the scheme, which only covered around 40 per cent of EU emissions. There are sectors that are even being overcompensated, which leads to windfall profits.[4] An overhaul of the EU's protection mechanism against this so-called 'carbon leakage'[i] is needed in order to ensure that the polluter-pays principle is respected. For the market to work properly there had to be scarcity, but so many permits were issued that the market collapsed twice, since there were more permits than people wanting to buy them—another consequence of fossil industry lobbying.

An academic analysis identifies another important design flaw: individual EU member states were allowed to set the quantity of permits issues, at least initially. The absence of a supranational enforcement agency led to both leakage from the cap and inefficient competition between the different countries.[5]

[3] https://ec.europa.eu/clima/policies/ets_en.

[4] Laing, Timothy, Sato, Misato, Grubb, Michael and Comberti, Claudia (2014), 'The Effects and Side-Effects of the EU Emissions Trading Scheme', *WIREs Climate Change*, 5/4, 509–519. https://doi.org/10.1002/wcc.283.

[5] Lapan, H.E. and Sikdar, S. (2019), 'Is Trade in Permits Good for the Environment?', *Environmental Resource Economics*, 72, 501–510. https://doi.org/10.1007/s10640-017-0202-z.

The Corporate Europe Observatory—a Brussels-based thinktank that assesses the negative impact of corporate lobbying on EU policymaking—goes much further, outlining five reasons why the scheme should be scrapped. According to their critique the ETS:

- **has not substantially reduced emissions:** GHG emissions have actually fallen because of other policies such as feed-in tariffs encouraging the growth of renewables, not because of ETS.
- is used to undermine other climate and emissions control policies: industry lobbyists have successfully used the ETS as an excuse to block other environmental regulations.
- **set a ceiling on climate ambition: this lack of ambition combined with the excess of permits has undermined the determination of national governments to tackle climate change.**
- has not been cost-effective and has subsidized polluters at taxpayers' expense, especially because of the allocation of free permits;
- **remains susceptible to fraud and gaming, especially because CO_2 is not a tangible product and the legal definition and time-period for emissions can both be gamed to allow leakage.**

As we will see in the following section, a proposal for a much stronger policy in the form of a carbon tax was defeated by exactly the same industry lobbying that has made the ETS ineffective.

Environmental Taxes

The other main policy proposal for pricing carbon is to introduce a form of carbon taxation While carbon trading has gained more media attention and rhetorical support, the initial and most obvious policy to reduce CO_2 emissions is to tax them. The most popular proposal is for a tax that is applied as a fuel tax, based on the amount of fuel sold. When the fossil fuel is burnt, CO_2 is released and the quantity is directly related to the amount of fossil fuel consumed. The tax could be imposed in a number of different ways. The simplest would be an upstream tax, imposed on oil and coal companies when they extract the fuel from the ground. This would ensure that the total quantity of fuel is taxed and would be simple and cheap to administer. It would then be the responsibility of the fuel companies to pass the cost on to intermediate producers, who would then in turn pass the cost on to consumers.

The immediate appeal of a system of taxation is that it would address all polluters, not just the businesses who would become part of a carbon trading system. Although taxation systems are costly to establish and monitor, they do not involve the transaction and negotiation costs that are present with any trading system. The advantage of a market system is that it would be self-adjusting, i.e. the price of a CO_2 permit would rise or fall according to demand. However, this could also be a significant disadvantage for businesses, because they would not be able to have a fixed idea about the cost of their emissions when producing business plans. There might also be a high degree of volatility in the price of CO_2 emissions, which could make planning difficult. A taxation system, by contrast, would be clear; it might be fixed on a clear and rising trend so that businesses could plan for the cost of fossil fuels to rise gradually over time, and they could factor this into their planning. Although such a cost would be unwelcome, it would at least be foreseen.

Perhaps the most attractive aspect of a taxation proposal is that it is a type of policy that is already familiar to both taxpayers and policymakers. Creating carbon markets, by contrast, is an innovative and highly complex process—the pitfalls of such an experimental approach have already been outlined above. A tax would also generate revenues that could be reinvested in the infrastructure of a low-carbon economy—being made available as grants for home insulation or transition grants for businesses to install renewable energy systems, for example. This apparent ‘benefit’ is something of a double-edged sword, however, since the public is sceptical about pro-environment taxes, which they suspect may be introduced primarily to generate revenue rather than to tackle the environmental problem.

It is also important to realize that a carbon tax will increase prices of energy but all goods that depend on energy, which means all goods! To prevent exacerbating social injustice it is essential that the revenue is reinvested to palliate these costs—say by providing cheap public transport of publicly funded home insultation schemes. Alternatively, the revenue could be distributed directly to the public as a form of citizens’ income as proposed by James Hansen in his fee-and-dividend scheme.[6]

[6] https://citizensclimatelobby.uk/wp-content/uploads/2021/05/FeeAndDividend.CliveEllsworth.July2014-1.pdf.

The idea of a tax on carbon, on fossil fuel use at point of extraction, was proposed by the European Commission back in 1992 but for political reasons rather than reasons of policy efficacy the Emissions Trading System was adopted instead. Clearly, if we had priced CO_2 emissions properly then we would not have found ourselves in a climate emergency. Various countries have experimented with certain types of carbon tax and with varying degrees of success.

In 2013, the UK introduced a 'carbon price floor', essentially a tax on the use of fossil fuels that made up the difference between the cost of CO2 permits under the ETS scheme and a fixed and rising trajectory for carbon emissions that the UK fixed to ensure it would meet its Kyoto commitments for carbon reductions. As the House of Commons Library explains[7]:

> When the CPF was introduced, it was due to rise every year until 2020 (to a price of £30/tCO2). At Budget 2014 the Government announced that the CPS component of the floor price would be capped at a maximum of £18/tCO2 from 2016 to 2020 to limit the competitive disadvantage faced by business and reduce energy bills for consumers. This price freeze was extended to 2021 in Budget 2016.

Essentially, a change of government to one with fossil fuel friends meant that the policy was undermined. It has yielded significant income—£1bn in 2016/7—but this was held by the central government and not invested in renewable energy or other sustainability measures. However, the policy has shifted UK electricity generation from a heavy dependence on coal towards the lower-emission gas fuel, as well as encouraging the spread of renewable generation.

The Australian economy is highly dependent on mining so the introduction of a carbon tax there in 2012 was always going to be controversial, although the price fixed at Aus$23 per tonne was offset by other tax benefits. A report by the Australian National University estimated that by 2014 the policy had led to the fastest ever fall in greenhouse gas emissions. But again, vested interests lobbied against the tax and later that year a conservative government repealed the law that put the tax in

[7] Hirst, David (2018), BRIEFING PAPER Number 05927, 8 January 2018, Carbon Price Floor (CPF) and the price support mechanism, Briefing Paper no. 05927 (London: House of Commons Library). https://researchbriefings.files.parliament.uk/documents/SN05927/SN05927.pdf.

place, meaning that Australia is no longer on track to meet its carbon commitments (Fig. 2).

It is clear that if some countries introduce a tax that increases their energy costs they will simply be exporting energy-intensive industries and jobs to countries that have not adopted such a policy. However, conversely, a powerful bloc like the EU can introduce a trade policy to ensure that imports bear the implicit cost of climate damage through a specific and targeted tariff and this can help to encourage countries who export to the EU to reduce their carbon intensity of production. In the European Union this is being discussed in terms of a 'carbon border adjustment', a tariff charged on imports from countries that are not addressing climate change to reflect the costs they have saved. This can be done through a mechanism that aims at adjusting import prices up to the level of EU's carbon price. The rules of the WTO, the international body established to ensure a level playing field for global trade, allow exemptions specifically related to environmental benefit and so long as domestic and imported products would be treated alike.

My strong support for a carbon tax—and indeed other environmental taxes—is that they work. Within a market system price signals change behaviour, and this is the biggest level we could use on a global basis. To quote a comment piece in the *FT*: 'Evidence on the effectiveness of carbon taxes is clear in the data. In sectors where environmental taxes have been becoming more onerous—industry, power generation and waste—emissions have been falling fast, whereas in sectors with stable taxation such as aviation, transport and agriculture, the improvements in carbon use have been much lower'.[8]

Allowing the market to price carbon is not working. As Kristalina Georgieva, boss of the IMF put it. 'this price signal needs to get predictably stronger—by 2030, we need an average global price of $75 per ton of CO2, way up from today's $3 per ton and up from the 23 percent current emissions coverage'.[9] Vested interests are blocking this essential tool to decarbonize the global economy and, if we do not resist

[8] Giles, Chris and Hook, Leslie (2020), 'Zero Emissions Goal: The Mess of Britain's Carbon Taxes', *Financial Times*, 10 March.

[9] Remarks by IMF Managing Director on Global Policies and Climate Change, International Conference on Climate, Venice, 11 July 2021. https://www.imf.org/en/News/Articles/2021/07/11/sp071121-md-on-global-policies-and-climate-change.

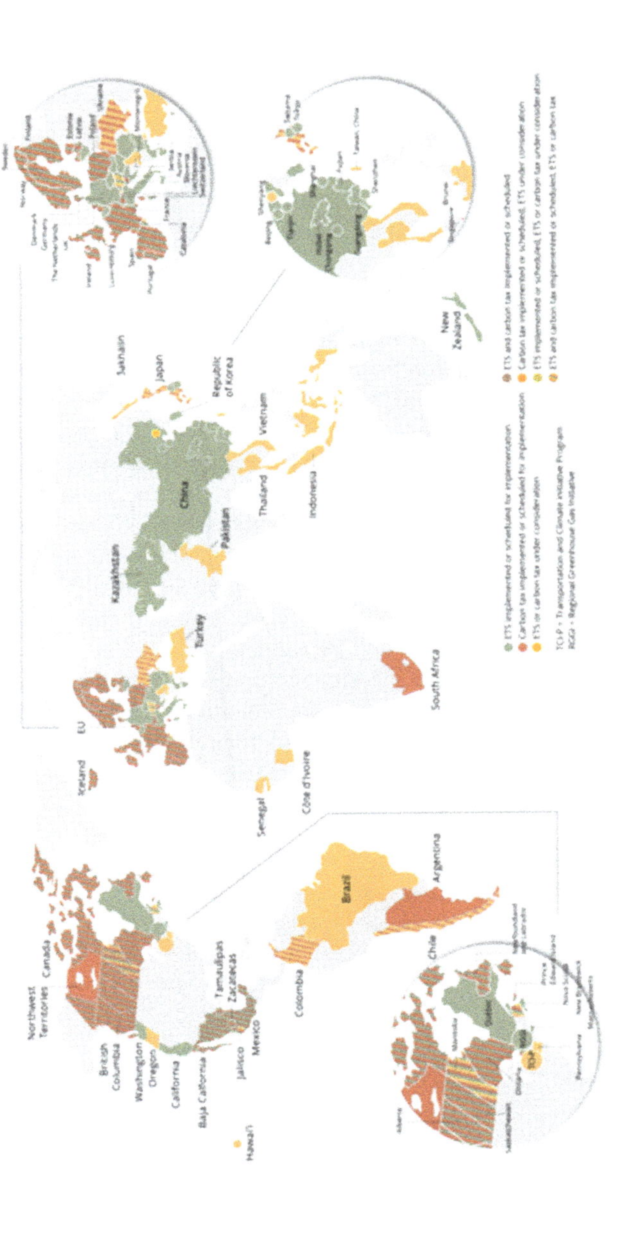

Fig. 2 Illustration of carbon tax and trading around the world (*Source* Based on World Bank data analysed by World Bank staff; thanks to Wikimedia Commons for making this graphic available free of charge [A/w available open source online here: https://en.wikipedia.org/wiki/Carbon_tax#/media/File:Carbon_taxes_and_emission_trading_worldwide_2019.svg])

their pressure, all our futures are at stake. The failure of the EU to introduce a carbon tax as long ago as 1992 is an object lesson is why human society stands on the brink of extinction not because of lack of scientific knowledge of policy innovation but because of the political power of fossil fuels.

3 Creating Sell-By Dates for Stranded Assets

In a paper based on our experience of European policymaking, Cory Fletcher and I discussed how the twin powers of sustainable finance regulation and banking oversight could provide a pincer movement to shift the European economy in the direction of a sustainable future.[10] Since EU policy-makers are responsible both for the assessment of the quality of financial assets, through their role as financial regulators and governers of financial stability and for the legislation that permits or bans different forms of economic activity, European policy-makers have the powers and tools they need to create this clear signalling framework. We proposed this as a process of creating and then eliminating stranded assets.

We considered this in the context of a vital sector whose unsustainable operating practices are eroding its own productive base and threatening its future: intensive agriculture, identified by the EU's High-Level Expert Group on Sustainable Finance as a priority sector for rapid acceleration of sustainable practices. HLEG identified risks including the degradation of soil quality; large-scale use of pesticides and their impact on natural resources (such as pollinators); and widespread use of antibiotics. They concluded that

> The current food production and supply system is complex, and it is largely opaque to both the end user and the financial investor....there is no mandatory disclosure on products or processes that were used in the production – such as pesticides or antibiotics. Information that is directly related to the sustainability of production methods should urgently be

[10] Scott Cato, Molly and Fletcher, Cory (2020), 'Introducing Sell-by Dates for Stranded Assets: Ensuring an Orderly Transition to a Sustainable Economy', *Journal of Sustainable Finance & Investment*, 10:4, 335–348.https://doi.org/10.1080/20430795.2019.1687206.

disclosed as it will enable the financial sector to encourage the transition towards a more sustainable model for agriculture.[11]

These unsustainable farming practices are not only contributing significantly to the ecological crises and devastating biodiversity, they are also undermining the profitability of the agricultural sector in the long term, thus creating stranded assets for those who have invested there.

Another area where legislation is likely to affect the value of assets is the extractive industries, as policy-makers enforce higher standards on mining companies, especially in terms of their activities in emerging markets. The appalling story of 'blood diamonds' or 'conflict diamonds'—extracted in appalling conditions, often by children—in countries such as the Congo and where the proceeds are used to finance devastating wars, has led to their being banned by many jewellers. A global campaign led to the passage of UN Security Council Resolution 1173 in 1998 and the Kimberley Process to monitor the supply chain for diamonds. The process has not been adhered to by African governments and has met limited success but it did inspire a similar process to exclude 'conflict minerals' from the EU single market.

The EU Conflict Minerals Regulation came into force on 1 January 2021. It requires that EU importers of 3TG metals (tin, tungsten, tantalum, and gold) must meet internationally responsible sourcing standards, set by the Organisation for Economic Cooperation and Development (OECD). From 2021, EU manufacturers of electronic goods using these four metals will have to monitor their supply chain to ensure they only import these minerals and metals from responsible and conflict-free sources. These minerals that are vital to a range of electronic consumer goods must be carefully monitored and certificated if the goods are to be sold to European consumers (Fig. 3).

The figure demonstrates how we might generalize this process that we referred to as 'setting sell-by dates for unsustainable assets'. The top bar in the graphic shows how the production of tablets and phones that currently use metals was in an amber zone once the Regulation was announced and has now moved into a red zone, meaning that the assets represented by investments in companies producing those devices have lost their value.

[11] HLEG (2018), 'Final Report 2018 by the High-Level Expert Group on Sustainable Finance'. Brussels: European Commission, p. 91.

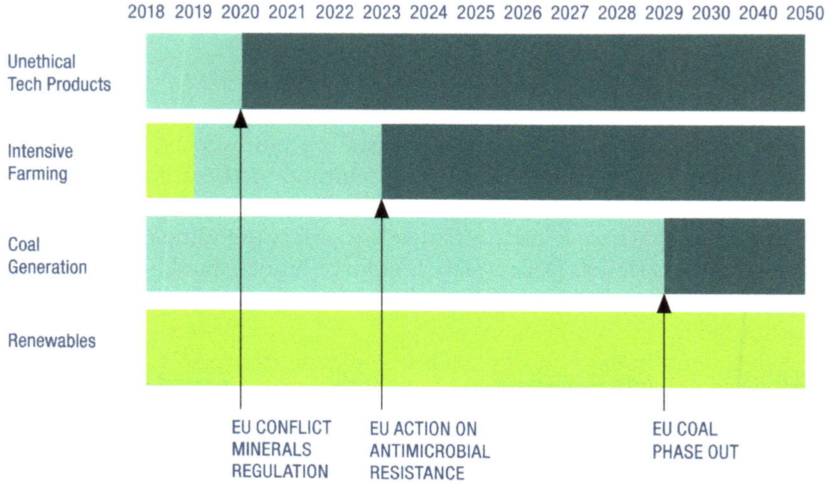

Fig. 3 A Timeline Depicting the Transition Period and Sell-by Dates for Stranded Assets According to EU Policy Commitments (*Source* Author's graphic redrawn by Angela Mak)

The second bar indicates how certain types of farms must respond to the restrictions on antibiotic use recently agreed upon in new legislation on Veterinary Medicinal Products (VMP) and Medicated Feed and that will come into force in 2022. Investment in highly intensive animal-rearing companies—that can only maintain such unhealthily high stocking densities by routine use of antibiotics that will soon become illegal—have moved from the green zone into the amber zone and will soon lose their value.

As for the two lower bars that represent the energy sector, the upper one illustrates the impending loss of value of Europe's coal-based generating capacity as the target date for net-zero generation hoves into view (dangerously late at 2030 but with a potential to be brought forward). The lower line represents renewable electricity generation which is already sustainable and so where investments have no prospect of becoming stranded.

In essence, what we proposed in our paper is a tool to facilitate the transition to a sustainable economy based on the sequential creation and elimination of stranded assets and a way to conceptualize the use of financial and environmental regulation to shift the European economy towards a sustainable future. Clear and early signalling should enable investors to shift their asset holdings to avoid a disorderly process that might risk financial collapse. Beginning with climate change, the concept of stranded assets can then be extended to a range of other environmental and social factors, based on the urgency of the ESG risk that they pose.

4 Saving Life on Earth: The Biodiversity Crisis

The focus of the sustainable finance debate is slanted towards climate for the obvious reason that it is the most urgent and devastating crisis we face as a human community. But we also face the much wider and deeper ecological crisis: our industrial and consumptive way of living is destroying the planet we depend on and creating an ongoing massacre of the species we share the planet with.

Economists were late to acknowledge this—we could even argue that the neoclassical economic model created a culture of denial that delayed action until it was too late for many precious species and ecosystems—but now they have come to the party they appear to be something of a mixed blessing.

This is exemplified by the Das Gupta review, commissioned by the UK Chancellor and published by the UK government in February 2021.[12] The headline message—that, as a human community, we are utterly dependent on the natural world—is one that I would wholeheartedly endorse as a green economist:

> We are part of Nature, not separate from it. We rely on Nature to provide us with food, water and shelter; regulate our climate and disease; maintain nutrient cycles and oxygen production; and provide us with spiritual fulfilment and opportunities for recreation and recuperation, which can enhance our health and well-being. We also use the planet as a sink for our waste products, such as carbon dioxide, plastics and other forms of waste, including pollution. Nature is therefore an asset, just as produced capital (roads, buildings and factories) and human capital (health, knowledge and skills) are assets.

[12] Dasgupta, P. (2021), *The Economics of Biodiversity: The Dasgupta Review* (London: HM Treasury).

But when this statement is moved into a discussion about how we should change our economic activity the agreement is not so clear-cut. Of course, Das Gupta is right that we have failed to work with and use the resources of Nature in a sustainable way. But is it helpful to frame the need for change in terms of extending the destructive economic logic into the natural world rather than learning the lessons of ecosystems and processes and exporting them to our design of the economy?

In the words of the Review:

> We are all asset managers. Individuals, businesses, governments and international organisations all manage assets through our spending and investment decisions. Collectively, however, we have failed to manage our global portfolio of assets sustainably. Estimates show that between 1992 and 2014, produced capital per person doubled, and human capital per person increased by about 13% globally; but the stock of natural capital per person declined by nearly 40%.

Some of the policy recommendations from the Das Gupta review are vital and urgent. They fall under three headings:

- Address the Imbalance Between Our Demand and Nature's Supply, and Increase Nature's Supply—conservation and restoration of ecosystems; reducing our demand on the natural world; questioning our level of consumption and the size of the global population; redesigning our systems of production and trade so that they are in balance with natural systems.
- Changing Our Measures of Economic Progress—end the focus on GDP as the primary measure of wealth and substitute more inclusive measures based on the well-being of people and planet; adopt national capital accounts so that losses borne by nature figure in national accounts.
- Transforming our Institutions and Systems—transforming our financial and educational systems; better protection of global public goods; redesign of the global financial systems; re-empowering of citizens.

Some of these propositions—particularly the need to abandon growth as the key metric of economic success and the welcome emphasis on the need to change consumption patterns—have been called for by environmentalists and green economists for decades. They are also clearly wide-ranging, underlining what we have always understood about the interconnected nature of the sustainability crisis. Those that focus particularly on the finance sector are of most interest here:

> Far more global support is needed for initiatives directed at enhancing the understanding and awareness among financial institutions of such Nature-related financial risks, learning and building on the advances on climate-related financial risks. Central banks and financial supervisors can support this by assessing the systemic extent of Nature-related financial risks. A set of global standards is required. They should be underpinned by data that are both credible and useful for decision-making. Businesses and financial institutions could then be obliged to integrate Nature-related considerations within their objectives. The idea ultimately is to have them assess and disclose their use of natural capital. The Task Force on Nature-related Financial Disclosures, established in 2020 [and that we will consider in the following chapter], is a step in that direction.

This quotation is helpful in terms of confirming that the main issues are those we have covered in this book. Essentially this confirms that in terms of analysis and disclosure of nature-related risks must be confirmed in the same way that climate-related risks increasingly are; that analysis and reporting of such risks is urgently necessary (as is required by the SFDR); that these risks needed to be incorporated into the policies of central banks.

It is undeniably true that our valuation system is entirely out of kilter in a world where, in the words of BBC Environment Editor Justin Rowlatt, the value of Amazon the global delivery is so much greater than the value of the Amazon rainforest, the lungs of the world. Or, in the words of Katie Kedward, a London-based economist, 'We have collectively failed to engage with Nature sustainably, to the extent that our demands far exceed its capacity to supply us with the goods and services we all rely on'.[13] But she also raised doubts about the ability of economists to estimate 'nature loss in monetary terms and given the uncertainties involved price-based tools can't be used as a replacement to more important steps to restore nature'.

[13] https://www.bbc.co.uk/news/science-environment-55893696.

This chimes with doubts raised by environmentalists ever since the whole idea of considering nature as 'capital' and the creation of the concept of 'ecosystem services' was first raised. For them, framing the environmental crisis in economic terms amounts to the commodification of nature and raises a number of problematic issues. The mainstreaming of this framing is encapsulated in the 2005 Millennium Ecosystem Assessment, conducted under the auspices of the UN.[14] The report's headline was that 'Over the past 50 years, humans have changed ecosystems more rapidly and extensively than in any comparable period of time in human history, largely to meet rapidly growing demands for food, fresh water, timber, fiber, and fuel. This has resulted in a substantial and largely irreversible loss in the diversity of life on Earth' (p. 2).

This is undeniable, but the question is whether using the language and theories of economics will help to reverse this destructive trajectory. Proponents of the concept of 'ecosystem services' freely admit that they have designed the concept to increase the attractiveness of talk of environmental protection to the corporate sector. If money talks, if the power of the world lies with those who control the economy, then better to speak this language than to appeal to people's innate desire to protect the natural world.

Academics have taken issue with this framing:

> What are 'Ecosystem Services'? At first hearing, they sound like a firm of consultants who help you repair your ailing ecosystem. In fact it's the other way round: the service is provided by people with ecosystems to people who no longer have one, and who need one. For example if your forest, or your peat bog is absorbing carbon, it is providing a service to other people who are producing excessive CO_2 and need something, somewhere to absorb it. Other ecosystem services include climate regulation, maintenance of biodiversity, water conservation and supply, and the preservation of aesthetic, cultural and spiritual values. The emerging view is that the people receiving these ecosystem services should start to pay for them. (Sullivan 2008a: 21)

Critics take issue with the concept on a number of grounds. First, it assumes that the nature and the creatures we share the planet with

[14] Millennium Ecosystem Assessment (2005), *Economy and Human Well-Being* (Washington: World Resources Institute). https://www.millenniumassessment.org/documents/document.356.aspx.pdf.

are here to service our desires; it puts humankind at the centre of the system, a fallacy referred to as 'anthropocentrism'. It also has an inevitable implication of instrumentality, meaning that nature is there to serve our purposes rather than having its own inherent value. The risk of taking this perspective is that we will seek to protect species that have a value for us—such as bees—but allow craneflies or moths to become extinct. Perhaps most importantly, from an ecological perspective, it is the whole ecological system—the web of life—that really matters. Different organisms are connected in complex ways that scientists still struggle to understand, let alone price. The commodification of parts of the natural world strips them of their inherent value and, perhaps most seriously of all, makes them at greater risk of exploitation. If we can cost a piece of rainforest accurately, why should we not allow somebody who can provide the Yanomami people who live there with a sufficient large volume of cash and insist that they leave?

There are also difficulties in terms of establishing ownership rights over the areas of the world where ecosystems remain intact, largely due to low levels of industrialization. Since the people living in these areas have less economic clout, they may not be well-placed to protect their rights over their land and their lifestyle—which, paradoxically, is precisely what has preserved the ecosystem. As environmental pressures increase, subsistence farmers in the world's poorer nations are threatened with displacement and loss of livelihood as their land is traded to provide carbon sinks and other ecosystem services for the peoples of the richer world. It may seem strange to think of the planet in terms of a range of services, and it seems to make little sense to attempt to measure them because their value is clearly infinite. Without the planet, we cannot survive, so all the money we could ever create would not be a large enough price to pay. More worryingly, perhaps if those who control finance could create enough of it they would be able to purchase what remains of these 'ecosystem services' and exclude others from them.

From a green economic perspective, this whole discussion would appear to suffer from the familiar problem of a category error, i.e. considering the ecosystem as part of the economy rather than the economy as an entirely dependent part of the ecosystem.

Note

i. https://ec.europa.eu/clima/policies/ets/allowances/leakage_en.

CHAPTER 5

Defining, Measuring, and Reporting Sustainability

Abstract Increasingly, reporting the bottom line—i.e. the finances—is not enough. This chapter explains the developing legislation requiring non-financial reporting, especially accounting for Environmental, Social, and Governance (ESG) impacts. This includes the growth in scope and complexity of private-sector guidelines for reporting as well as consideration of the UN Taskforce on Carbon-Related Financial Disclosure and the new EU mandatory disclosure regime. Finally, the chapter looks at the issue of greenwashing and how the EU taxonomy might be able to define what is sustainable in a way that limits or eliminates greenwashing.

Keywords Non-financial reporting · ESG reporting · Mandatory disclosure · TCFD · Greenwashing · EU taxonomy

1 Non-financial Reporting: Beyond the Bottom Line

What goes on inside the black box of companies? Most of the legal duties on companies relate to them providing returns to shareholders and reporting their financial situation in an honest way. It is, to use an overused cliché, all about the bottom line. Part of the sustainability agenda arises from investors, regulators, and citizens expecting more from

© The Author(s), under exclusive license to Springer Nature Switzerland AG 2022

M. Scott Cato, *Sustainable Finance*,
https://doi.org/10.1007/978-3-030-91578-0_5

companies and demanding to know about how their money is made and spent, rather than just how much of it there is. This area is known as reporting and disclosure and it is vital both to coopt accountants into the sustainability effort but also to prevent greenwashing. Without clear reporting on environmental and social risks and benefits, investors will not have clear signals about where to send their money and managers will not have clear information about business opportunities.

According to a report commissioned by the UK's business department, 'Non-financial reporting refers to reporting on any matters relating to activities of a business that are beyond the financial transactions and financial standing of a business'.[1] While there are accounting standards for financial reporting there are, as yet, no internationally accepted standards for non-financial reporting. But there is pressure for change arising from a number of sources identified in the report:

- Environmental concerns, including the role that businesses and others can play in the
- transition to a low carbon economy.
- Consumer interest in sustainable investment, where investors wish to earn a return but
- 'do no harm'.
- A perception that to rebuild trust in business, businesses need to demonstrate a wider
- purpose beyond making profit.
- Trends towards greater litigation as individuals and non-governmental organisations
- (NGOs) seek to hold businesses to account for their actions and impacts.

There is clearly a need to standardize non-financial reporting standards and the European Commission has shown leadership here in terms of its Non-Financial Reporting Directive.

[1] Eunomia (2020), 'Frameworks for Standards for Non-financial Reporting: Final Report', BEIS research paper number 2020/052 (London: BEIS).

The EU launched its Non-Financial Reporting Directive NFRD in 2018, for the first time imposing obligations on large European companies to disclose information on their sustainability risks and impacts.[2] While this was an important first step, the Directive has been criticized for lack of stringency around the quality and relevance of information required. The EU Commission has itself accepted that the way information was reported under the Directive was not easily comparable from company to company, missed out some vital elements, and was not widely trusted as reliable.

The EU is now revising this Directive and renaming it as the Corporate Sustainability Reporting Directive, a key project of the Commissioner for Financial Services Mairead McGuinness. In her words, 'We need to mobilise at least half a trillion euros per year of additional investments in the EU' and sustainability reporting is 'one of the priorities to strengthen the foundations for sustainable investment'.[3]

The proposal for the new law includes the following strengthened requirements[4]:

- extends the scope to all large companies and all companies listed on regulated markets (except listed micro-enterprises)
- requires the audit (assurance) of reported information
- introduces more detailed reporting requirements, and a requirement to report according to mandatory EU sustainability reporting standards
- requires companies to digitally 'tag' the reported information, so it is machine readable and feeds into the European single access point envisaged in the capital markets union action plan

[2] European Commission: Non-financial Reporting: EU Rules Require Large Companies to Publish Regular Reports on the Social and Environmental Impact of Their Activities. https://ec.europa.eu/info/business-economy-euro/company-reporting-and-auditing/company-reporting/non-financial-reporting_en. Last retrieved: March 15, 2021.

[3] 'Reform of the EU Non-financial Reporting Directive: A Push Towards Future-proof Reporting Obligations', Full Disclosure: Monthly Briefing on the EU Corporate Transparency Regulation, March 2021. https://germanwatch.org/sites/default/files/Full%20Disclosure%202021-3_Reform%20of%20the%20EU%20Non-financial%20Reporting%20Directive.pdf.

[4] https://ec.europa.eu/info/publications/210421-sustainable-finance-communication_en.

The original NFRD began with Europe's largest companies—those with more than 500 employees and that were listed on stock exchanges.[5] This meant that only 11,000 companies were required to report. The revised reporting law will include smaller companies, although SMEs will still be exempt. This strengthened reporting regime for the EU single market goes hand in hand with the mandatory disclosure legislation (SFDR) that I discuss in detail in Sect. 3.

While the EU is ensuring that its own regime for disclosure of ESG risk is strong and coherent, policy-makers are keen to see it fit within a global framework of shared standards. This means coherence with initiatives by the G20 and G7, the various accountancy and financial stability organizations operating across the world, and the TCFD (Task Force on Climate-related Financial Disclosures), of which more shortly.

2 Private Sector Leaders

The previous section betrays my bias towards policymaking and my conviction that strong regulation is needed to make sure that companies—including finance companies—do the right thing. But there were actually important steps taken towards better transparency and reporting by organizations taken by private companies themselves before politicians obliged them to take action.

Nick Robins, then CEO of HSBC's Climate Change Centre of Excellence and later head of the UNEP work on sustainable finance and a co-founder of Carbon Tracker, described in 2014 the rapid evolution of action among private financial institutions who had expert staff dedicated to the sustainability agenda. He identified three types of initiatives:

- Disclosure, i.e. the provision of information to investors beginning with carbon disclosure but moving on to disclosure of Environmental, Social, and Governance impacts of the firm's investments.
- Responsibility, beginning with voluntary Principles for Responsible Investment, reporting these as part of annual accounts and using them to influence investment decisions.

[5] EU Commission (2021), 'Questions and Answers: Corporate Sustainability Reporting Directive proposal'. https://ec.europa.eu/commission/presscorner/detail/en/QANDA_21_1806.

- Climate Change, including a particular focus on climate-positive investments and lobbying governments and policy-makers to introduce policies to reduce climate risks.

An academic analysis of non-financial reporting frameworks identifies three waves of their development[6]:

- The first wave dates back to the 1970s, was led by the OECD—especially its Guidelines for Multinational Enterprises—and emphasized in particular employment and industrial relations standards while neglecting the environment.
- The second wave ran through the 1990s and into the early twenty-first century and was identified with the Eco-Management and Audit Scheme (EMAS) (1993), the Global Reporting Initiative (1996); the UN Global Compact (UNGP) (2000); and the Carbon Disclosure Project (CDP) (2000). The rhetoric during this wave was focused on 'social responsibility' and voluntary compliance rather than regulation and we can see the growing emphasis on climate and environmental issues.
- We are now in the third wave, which the authors date back to the 2008 financial crisis and the perceived need for stronger regulation, especially of finance. They note a number of national legislative initiatives focused on business responsibility including the Dodd-Frank Act in the US (2010); the 'Devoir de Vigilance' Law in France (2010); and the Companies Act in the UK (2013). They also note the tightening of the OECD Guidelines on Business Responsibility and the revision and expansion of scope of the GRI.

The authors conclude that there is a move away from reporting as a form of corporate communication and towards greater accountability and supply-chain management. They also note a growing emphasis

[6] Monciardini, David, Mähönen, Jukka Tapio and Tsagas, Georgina. (2020), 'Rethinking Non-Financial Reporting: A Blueprint for Structural Regulatory Changes', *Accounting, Economics, and Law: A Convivium*, 10:2. 20,200,092. https://doi.org/10.1515/ael-2020-0092.

on the link between reporting and sustainable investment, focused on Environmental, Social, and Governance (ESG) impacts.[7]
The highest profile initiative in this space is the Taskforce on Climate-related Financial Disclosures, set up by the Financial Stability Board of the Bank for International Settlements under the leadership of Michael Bloomberg and at the request of the Chairman of the BIS Mark Carney. It is important to note the provenance of this organization in capital markets, so serving the interests of investors rather than citizens, and its location in the private sector, so the industry regulating itself, rather than being a form of democratic regulation, which is the focus of EU activity discussed later. It is also limited to *climate* risks to investments rather than the wider ESG agenda.

Nonetheless, the TCFD has shown important leadership and, due to its provenance, has acquired the confidence of many financial-market agents and companies. Its final report, published in 2017, states clearly that the risks to the value of assets from climate change have been underestimated and that this threatens the stability of financial markets[8]:

> The reduction in greenhouse gas emissions implies movement away from fossil fuel energy and related physical assets. This coupled with rapidly declining costs and increased deployment of clean and energy-efficient technologies could have significant, near-term financial implications for organizations dependent on extracting, producing, and using coal, oil, and natural gas.

The report quotes a previous study that estimates this risk as ranging from \$4.2 trillion to \$43 trillion between 2015 and the end of the century. The 32 members of the Taskforce were drawn from across the world and represent the relevant industry sectors: banks, insurance companies and pensions funds, assets managers, accountants, credit-rating agencies, and including non-financial companies.

[7] For those interested in the pre-history of non-financial reporting, see: Carlos Larrinaga and Jan Bebbington (2021), 'The Pre-history of sustAinability Reporting: A Constructivist Reading', *Accounting, Auditing & Accountability Journal* (May). https://www.emerald.com/insight/content/doi/10.1108/AAAJ-03-2017-2872/full/html.

[8] TCFD (2017), 'Recommendations of the Task Force on Climate-related Financial Disclosures'. https://assets.bbhub.io/company/sites/60/2020/10/FINAL-2017-TCFD-Report-11052018.pdf.

The role of the Taskforce was to develop consistent standards for financial-market players to use in reporting climate-related risks. The Taskforce's recommendations apply to financial-sector organizations, including banks, insurance companies, asset managers, and asset owners. They boil down to three essential points:

- Carbon-related risks should be reported in mainstream financial filings, since they constitute a material risk to company valuation.
- They should be reported according to four elements, building outwards: metrics and targets; risk management; strategy; and governance.
- Such disclosures should take into account a range of future climate scenarios including a 2° Celsius climate warming or lower scenario.

Given that Mark Carney was Governor of the Bank of England when he began leading the debate on climate-related risks to financial stability, it is unsurprising that the UK government has responded positively to these recommendations. As part of its Green Finance Strategy, announced in 2019, the UK Treasury established its own Taskforce to explore the most effective response to the TCFD recommendations.[9] In November 2020 it announced an intention to make the TCFD climate-disclosure standards mandatory across the UK economy by 2025. EU regulators are taking a more inclusive and socially focused approach to the issue of non-financial reporting, as we will see below, but they have stated a determination to work cooperatively with the private-sector guidelines developed by the TCFD.

There is insufficient space here to follow up on all the key players identified in this field but a few merit particular attention.

The Global Reporting Initiative, based in the Netherlands, has shown important leadership in this area.[10] Describing itself as 'the independent, international organization that helps businesses and other organizations take responsibility for their impacts, by providing them with the global common language to communicate those impacts', it first launched in

[9] HM Treasury (2020), *A ROADMAP Towards Mandatory Climate-Related Disclosures* (London: HMT): https://assets.publishing.service.gov.uk/government/uploads/system/uploads/attachment_data/file/933783/FINAL_TCFD_ROADMAP.pdf.

[10] https://www.globalreporting.org/.

1997 in the US and with UN support, it formally launched its voluntary reporting standards in 2000 and has updated them regularly since then. They provide a framework for reporting by responsible companies and offer a comparable means of disclosing performance indicators in terms of climate, environmental, and social impacts.

Now that accountants are being required to report on non-financial matters they need to have clear frameworks within which to do this. International Financial Reporting Standards (IFRS) is a not-for-profit organization that was established to develop a globally accepted set of accounting standards and is now extending its activity in the area of sustainability.[11] Accounting standards are important because they enable those wishing to invest in companies to have clear and accessible information about how the companies are performing; they are legally required to be published in annual reports. In April 2021, the IFRS made moves towards establishing an International Sustainability Standards Board as the first step to develop accounting standards explicitly relating to sustainability.

With the increased focus on sustainability reporting we are also seeing changes in the voluntary bodies that have developed the idea of non-financial reporting, alongside the mainstream accounting bodies making moves in this direction. The International Integrated Reporting Council (IIRC) and Sustainability Accounting Standards Board (SASB) have merged into the Value Reporting Foundation. They are working with three other organizations that have pioneered ESG reporting—the Carbon Disclosure Project, the Climate Disclosure Standards Board, and the Global Reporting Initiative (GRI), to become the Group of Five. Their intention is to produce a disclosure standard focused specifically on climate risks: this 'sustainability-related financial disclosure standards would enable disclosure of how sustainability matters create or erode enterprise value'.[12]

Since this section has mostly been about how companies explain the social and environmental impacts of their activities I should probably end by explaining how this connects to the sustainable finance agenda. As I explained in Chapter 2, whether you have a pension fund, insurance, or

[11] https://www.ifrs.org/projects/work-plan/sustainability-reporting/.

[12] Fishman, A. (2021), 'What You Should Know About potential New International Reporting Standards', *GreenBiz*, 19 July. https://www.greenbiz.com/article/what-you-should-know-about-potential-new-international-reporting-standards.

just a bank account, financial intermediaries will be using your money in the world. It is easy to imagine it sitting in a bank vault somewhere but it isn't: it is out and about and if it isn't doing good stuff it is probably causing damage. Your money might have bought shares in companies or been loaned to companies—so you need to know what those companies are doing—that's where the reporting by companies comes in. In the following section we go on to consider how the financiers are now being required to report about the way they invest your money in other companies, and how they report depends on how comprehensively and reliably the companies themselves report their activities.

3 How Much Do You Know About What Your Money is Doing?

At the level of retail banking and financial services we all have a right to know what the banks are doing with our money. That may seem an obvious statement but I was astonished to find it was controversial and a principle I had to fight for as an MEP. When people found there was horsemeat rather than beef in their lasagne there was an outcry, but when the equivalent happens in the financial sector customers don't even have the right to know. I am pleased to say that, after a lengthy political struggle in the EU institutions, the reporting requirements for financial players are much stronger. What I can guarantee you was a hard-fought and intensely negotiated piece of legislation came into the public domain in March 2021 with the rather unappealing label of 'mandatory disclosure' and the abbreviation SFDR.

SFDR specifies the requirements for investment companies, pension funds, insurance companies and banks that provide investment services, to disclose how they take the environmental or social impacts of their investments into account. With a few limitations we were forced to concede during negotiations, it requires all large financial-market players (those with more than 500 employees) to gather and make public this detailed information about the impacts of their investments. Any firm that markets its financial products as 'sustainable' faces even stricter disclosure requirements to prevent greenwashing.

The SFDR is the world's first mandatory disclosure regime for a major financial market. The regulation includes a 'comply or explain' exemption, meaning that larger companies can choose not to reveal the ESG impacts of their investments for the first year of the regulation (until 30 June

2021). From that date all large financial-market players (those with more than 500 employees) will be mandated to comply. Any firm that markets its financial products as 'sustainable' will need to disclose the ESG impacts of the companies it invests in as soon as the regulation comes into force.

Integration of ESG factors into the financial framework is needed in order to reveal which investments are being directed towards projects that undermine the quality of the environment and social well-being. As a member of the European Parliament's Latin America delegation, I met survivors of the Fundão dam disaster in Mariana, Brazil. This disaster is a classic example of the damage to people's lives when corporations and the banks who finance their activities are not adequately regulated. HSBC and BNP Paribas finance Samarco Mineração SA (Samarco) mining company yet many of those who have accounts or savings with these banks did not know their money was being invested in such a potentially destructive way.

Financial products are now defined according to their level of pro-sustainability performance. Rather unromantically, these have acquired labels according to the paragraph of the EU regulation that defines them! These are:

- Article 6: Investments that show the minimum commitment to sustainability and provide a basic assessment and information about their ESG impact;
- Article 8: Investments that are considered to promote environmental and social benefit via their economic activity and take ESG criteria into account as part of their investment decision-making;
- Article 9: Investments that specifically aim to achieve sustainability outcomes alongside financial returns. They report their ESG performance on climate, environment, and wider social and governance agendas including human and worker rights.

According to the *FT*[13]: 'The new rules could have far-reaching consequences for asset managers—not just in Europe but around the world as investment firms are forced to demonstrate they are serious about sustainability. They will also influence the decisions of listed companies which

[13] Attracta Mooney (2021), 'Greenwashing in Finance: Europe's Push to Police ESG Investing', *FT*, 10 March.

will find themselves under pressure to focus more on ESG issues or risk losing investor capital'.

It is as yet unclear whether this mandatory disclosure regime will be followed by UK finance houses that now operate outside the EU regulatory area. Although the SFDR was not transferred into UK law at the end of the Brexit transition period, UK financial firms will need to meet its requirements if they market funds into the EU or act as a delegated investment manager to an EU firm. It may not be possible for them to resist, given the increasing emphasis on tighter regulation and growing demand for ESG monitoring by clients. After years when investment companies and pension funds had a standalone sustainability officer, the sustainable finance agenda is now being integrated into all investment decisions, with an assessment of ESG performance being put at the heart of portfolio management. The expectation is that UK financial firms will choose to comply on a voluntary basis to meet the global gold standard for disclosure.

4 Greenwashing

As investors and their intermediaries indicate their support for pro-sustainability investment choices, the risk of greenwashing increases. As customers we need to be sure that something that claims to be a green financial product is really green: we need to be able to have a means to distinguish the green from the merely greenwashed.

Greenwashing is familiar to those of us who have sought to shift our consumption in an environmentally friendly direction. The claims made by companies for the green credentials of their firms or products have become an epidemic—if even an oil company can claim to be part of our sustainable future because it is 'beyond Petroleum' or our government can claim that Jet Zero can make aviation sustainable, then the very concept of 'green' loses all meaning. While some of this PR activity is farcical and can easily be dismissed as corporate spin, it matters because if we cannot be sure that what is labelled as green is really green why should we not just return to unsustainable consumption habits?

An academic survey of definitions of greenwashing over the past decade confirms that this epidemic is well advanced.[14] They suggest a number of different forms of greenwashing:

- Greenwashing as selective disclosure: an egregious example from my neighbourhood is Bristol Airport claiming it will achieve Net Zero by 2030 but only because it will not include the emissions from flights in its calculations[15];
- Greenwashing as decoupling: companies focusing on their corporate responsibility in one area to distract from poor performance elsewhere, for example the community donations by supermarket chains that nonetheless have energy-intensive supply chains, use offers that lead to food waste, and a range of other environmentally destructive practices;
- Greenwashing as signalling: here the company makes no attempt to change its actual behaviour but rather uses subliminal signals of its intent, such as a product being displayed in a field or forest (Fig. 1).

The EU has been working for some years to produce a clear way of defining what is sustainable economic activity and what is not, using the somewhat scientistic word 'Taxonomy' to define this work. The taxonomy defines six types of economic activity to be sustainable:

- Climate change mitigation
- Climate change adaptation
- The sustainable use and protection of water and marine resources
- The transition to a circular economy
- Pollution prevention and control
- The protection and restoration of biodiversity and ecosystems

Exactly how these are defined relies on a great deal of technical work from the Commission's Technical Expert Group, who published their detailed

[14] de Freitas Netto, S.V., Sobral, M.F.F., Ribeiro, A.R.B. et al. (2020), 'Concepts and Forms of Greenwashing: a Systematic Review'. *Environ Sci Eur* 3219. https://doi.org/10.1186/s12302-020-0300-3.

[15] Estel Farrell Roig (2021), 'Bristol Airport ACCUSED of 'greenwashing' by Green councillor After Net Zero Pledge', *Bristol Post*, 28 June.

Fig. 1 The greenwashing photo opportunity (*Source* Cartoon by Timo Essner of the Cartoon Movement)

report final report on EU taxonomy On 9 March 2020. The taxonomy covers 13 sectors, including renewable energy, transport, forestry, manufacturing, buildings, insurance, and even the arts, which together account for nearly 80% of EU greenhouse gas emissions The hope is that the taxonomy can be used as the basis for a number of future policies including a green investment label and for EU-accredited green bonds.

Launching the legislation in April 2021, EU Commission Vice-President Valdis Dombrovskis said 'The taxonomy describes which economic activities are in line with the Paris Agreement, helping companies and investors to know whether their investments and activities are really green... it sorts green from greenwash'. While this is a laudable ambition it is clear that the taxonomy has come under considerable political pressure during its negotiation. No agreement was reached as to whether gas and nuclear should be included in the taxonomy and so this question was left to be decided later. The fossil fuel industry will be pleased that their attempt to portray gas as a 'transition fuel' has not been entirely dismissed.

The other controversial area concerns land use and bioenergy. This is another major political battle and in this case linked to the highly controversial area of reform of the EU Common Agricultural Policy. Although claims are made by some member countries that they have a sustainable forestry sector, including bioenergy in the taxonomy will mean encouraging the burning of trees that are needed as part of land-based carbon sinks.[16]

A critique of the EU taxonomy—a friendly critique that is intended to learn from weaknesses to strengthen a similar taxonomy set up by the UK government—identifies three weaknesses[17]:

- The EU Taxonomy is a positive list and needs to have a parallel list of unsustainable activities to prevent the creation and exploitation of loopholes.
- The definition of what is 'green' is too broad to have the necessary pro-sustainability impact, in particular including a number of industrial activities that are carbon-intensive but have the potential to undergo a transition to lower intensity.
- The binary nature of the Taxonomy means that companies in certain sectors that have a very different environmental performance are all either in or out.

[16] Frédéric Simon (2021), 'EU Spells Out Criteria for Green Investment in New "Taxonomy" Rules', *Euractiv*, 21 April. https://www.euractiv.com/section/energy-environment/news/eu-spells-out-criteria-for-green-investment-in-new-taxonomy-rules/.

[17] Yannis Dafermos, Daniela Gabor, Maria Nikolaidi and Frank van Lerven (2021), 'Greening the UK FINANCIAL system—A Fit for Purpose Approach', *SUERF Policy Note No*, 226:9. https://www.suerf.org/docx/f_55c6017b10a9755ef3681b09ccb01e94_21233_suerf.pdf.

Point 2 is open to debate, since accelerating the sustainability transition requires exerting pressure on energy-intensive industries to rapidly reduce their CO_2 emissions. The key to getting this right is identifying those sectors which will play a part in the future sustainable economy but need to radically transform and incentivizing them to do that. For example, we will always need steel but we need finance to flow to companies that either recycle it or produce it with a fraction of the energy of existing processes. Simply starving the steel industry of funds because it is currently 'unsustainable' might hold back the transition. The authors are right to point to the risk of greenwashing, but that must be avoided by insisting on scientifically verifiable evidence of pro-sustainability movement. And it is also vital to distinguish companies and industries that fit within the sustainability taxonomy, from those that can only ever improve their performance but never claim to be part of the green economy.

Extending this criticism, Reclaim Finance were much more trenchant in their critique, outlining how the gas and nuclear industries are exploiting the definitional weaknesses to undermine the impact of the Taxonomy[18]:

> When is a sustainable taxonomy not sustainable? When it becomes captured by lobbyists from environmentally damaging industries, of course. That is the key – unsurprising but unfortunate – takeaway of our new report on the gas and nuclear industries' €85m lobbying offensive aimed at overturning their initial effective exclusion from the European Union's flagship 'sustainable taxonomy', with support from state backers. Unfortunately, it appears their efforts are paying off, with the door now open to the inclusion of these energy sources confirmed in the European Commission's "sustainable finance strategy".

Their research shows that 182 gas-related entities spent between €64.9 and €78.4 million annually lobbying on the taxonomy. Between January 2020 to May 2021, the crucial period when the technical standards for the taxonomy were being finalized, gas lobbyists held an extraordinary 323 meetings with EU officials—more than one meeting every two days—including 27 meetings (8%) concerning the EU taxonomy or sustainable

[18] Reclaim Finance (2021), Out with the Science, in with the Lobbyists: Gas, Nuclear and the EU Taxonomy. https://www.Report-EU-taxonomy-Out-with-science-in-with-lobbyists-RF.pdf(reclaimfinance.org).

finance strategy. As for the nuclear industry, they found that 27 organizations are spending between €6.3 m and €7.9 m a year on pro-nuclear activity and that during the same period they held 44 meetings with the Commission, nine of which were on the topic of the EU taxonomy or sustainable finance strategy.

In all these discussions, the question is whether you limit sustainable finance to the deep green sectors or use it to accelerate the sustainability transition. The purpose of the taxonomy is to define the deep green sectors that need to be rapidly developed to accelerate the sustainability transition. That doesn't mean there is no role for companies that can rapidly transition to make themselves sustainable in the future economy. But there are sectors and companies that can never be sustainable. The first question must be: does this industry have a place in a sustainable future? Since there is no such thing as a sustainable fossil fuel it is absurd to include gas in a sustainable taxonomy. And it is self-defeating, since if the taxonomy itself becomes a form of greenwashing it will lose credibility and no longer be used as a standard against which to measure green investments.

5 The Beginning of a Journey Towards Sustainability

Sustainable Finance is moving rapidly in many directions and the issue of how and which companies should report is a rapidly moving field, as this chapter has demonstrated. Among regulators it is clear that the EU is showing leadership but the UN has also played an important facilitating role, especially in the area of climate. And there are numerous non-governmental organizations working to develop strong and trusted reporting standards. We can only assume—and indeed hope—that the rather mixed picture will resolve itself over the next few years with a positive collaboration between policy-makers, accountants and their professional bodies, and independent standard-setting bodies to achieve a single set of global standards to make clear information about ESG risks to shareholders and investors.

Although this chapter is focused on *global* reporting standards it seems to be slanted towards EU legislation and frameworks. That is clearly partly because of where I am personally situated but it is also because the US does seem to have lagged behind on this agenda. An academic analysis suggests this is the result of both corporate resistance and institutional barriers[19]:

> In general, the institutional foundations of emerging non-financial reporting and sustainability-oriented polices outside the U.S. include an emphasis on stakeholder interests, a comfort with the government's guiding role in the economy, and a far greater reliance on flexible principles-based disclosure approaches and on regulatory enforcement rather than on prescriptive rules and enforcement through shareholder litigation However, all of these stand in stark contrast to the norms and practices that shape the perspectives of U.S. regulators, legislators, and reporting companies themselves, and which constrain the scope and pace of non-financial reporting reform in the U.S.

The first priority is a clear and unified reporting framework for reporting non-financial impacts of economic activity. It is important that this is relevant and credible at a global level, because that is the level at which so much of our economy is now governed. But it is also vital that political regulators ensure that sustainable finance not only maintains its credibility and authority but also responds to social needs. If policy-makers cave in to corporate pressure then any number of EU kitemarks or green labels will descend into the farce of greenwashing. But if policy-makers keep their nerve then reporting and labelling could become a means whereby, step by step, we eliminate from the global economy products, activities, and companies that are damaging climate and ecosystems and violating human rights. This is the noble aspiration that lies at the heart of the rather unpromising agenda of 'non-financial reporting'.

The taxonomy and climate-related disclosures are just the beginning: what we need to be asking is the question asked by two academics in their paper 'How green is green enough?'.[20] Focusing on the political battle between the Parliament and the Commission over the Taxonomy proposal, they note that we need to move beyond the binary definition

[19] Harper Ho, Virginia. (2020), 'Non-Financial Reporting & Corporate Governance: Explaining American Divergence & Its Implications for Disclosure Reform', *Accounting, Economics, and Law: A Convivium*, 10:2, 20,180,043. https://doi.org/10.1515/ael-2018-0043.

[20] Trippel, E. (2020), 'How Green is Green Enough? The Changing Landscape of Financing a Sustainable European Economy'. *ERA Forum*, 21:155–170. https://doi.org/10.1007/s12027-020-00611-z.

of 'green' or not or 'green' or 'brown' and towards a 'shades of green' approach. And the transifigurtion to sustainability will mean that what is considered green this year, will not meet the tightened requirements next year or the year after. The Parliament was also determined to remove the loophole in the Taxonomy—that there is no brown taxonomy. Here is where the lobbying is fiercest but where the policy-makers need to have the courage to be clear about which industries do not have a place in the sustainable economy of the future.

We can imagine a future where, before you buy a winter coat or an electric bike, you can check the ESG label to see how the product scores in terms of the treatment of the people who made it or the amount of energy used or pollution emitted. As you now check your washing machine for an energy-efficiency market you can have access to summarized but reliable ESG data about everything in your life. But this will only work if the information is scientifically measured and honestly reported and that relies on political commitment and requires that our politicians resist the lobbying of those who will seek loopholes and opportunities to water down this vital regulation.

CHAPTER 6

The Role of Central and Public Banks

Abstract The chapter begins by considering why, in a climate emergency, banks are still investing in fossil fuels, and provides an account of who are the big players in this dirty market. Central Banks have been operating until recently within extremely restricted mandates, largely focused on low inflation and with an eye on unemployment. Growing demands for the sustainability transition to become part of the mandate of central banks were answered for the Bank of England in the budget in March, setting the pace for other countries. This chapter will also cover the issues of carbon stress tests and credit guidance, as well as exploring the role of public development banks such as the German KfW in financing the sustainability transition. It ends with a brief note about the role of the world's bank, the IMF.

Keywords Basel framework · Capital reserves · Central banks · Green QE · Credit guidance · KfW · IMF

© The Author(s), under exclusive license to Springer Nature Switzerland AG 2022
M. Scott Cato, *Sustainable Finance*,
https://doi.org/10.1007/978-3-030-91578-0_6

1 What Is the Role of Banks in a Climate and Ecological Emergency?

While almost all of us have regular interactions with banks for many of us the understanding of the role they play in our economy is rather hazy. In this chapter, I hope to open up the black box of banking and explore both the role banks have played in funding the dangerous industries that are driving us to environmental destruction but also how banks—and the politicians who regulate them—can contribute to the sustainability transition.

The reason sustainable finance has become such a focus in recent years is that, without finance, nothing will happen in the global economy. So finding ways to incentivize lenders to prioritize the sectors and industries that can accelerate the move towards sustainability is a powerful tool. But banks are also responsible for creating money and the role of central banks in controlling the money supply of individual countries is vital. So far they have played a limited role in directing the money supply so we will explore some ways in which they could play a stronger role. We also need to consider the potential of the world's global banking institutions—the World Bank and International Monetary Fund—and question why they haven't done more to create the finance the world needs to avoid environmental disaster.

Then we have public banks that are operating explicitly for the public good and do not need to focus on maximizing returns. Some of these public banks and development banks have led on financing the transition and played a vital role in providing finance for sectors that might otherwise struggle to find funding.

2 Banking Bad

In earlier chapters I have applauded the pensions and insurance industries for showing leadership on the sustainable finance agenda. The banks have done the reverse; incredibly, in the middle of a climate emergency, they are still lending far more to the dirty industries that we need to leave behind than to the green industries that represent our only hope of a future.

In March 2021, a group of NGOs including BankTrack, Indigenous Environmental Network, Oil Change International, Rainforest Action Network, Reclaim Finance and the Sierra Club published the results of their latest study into the extent of investment in fossil fuels by the 60 largest commercial and investment banks.[1] Their headline finding is that the banks invested $3.8trn in fossil fuels between 2016 and 2020 and that the level of investment was higher in 2020—even against the background of the pandemic—than it had been in 2016. They call out in particular JPMorgan Chase as the world's worst offender followed by Citi among US banks and Barclays as the European bank with the worst record. The following examples of dirty fossil fuel investments feature prominently on the wall of shame:

Tar sands oil: 2016–2020 financing was dominated by the Canadian banks, led by TD and RBC, as well as JPMorgan Chase. Because of the chemical nature of the fuel and because of where it is located, tar sands represent one of the most inefficient forms of fossil fuel taking nearly as much energy to extract as it produces. It also decimates the local environment where it is extracted and, in the case of the Canadian tar sands, this has violated indigenous people's rights.

Arctic oil and gas: JPMorgan Chase, ICBC, China Minsheng Bank, and Sberbank are the biggest funders since Paris, happy to support drilling in this highly sensitive region even as we see unprecedented melting.

Offshore oil and gas: BNP emerged as the world's worst banker of offshore oil and gas over the last five years. The eyes of the world turned to the Cambo field off Shetland, Scotland, as the British government havers over whether to grant permission for oil extraction there while also leading the COP26 negotiations.

Fracked oil and gas: US banks like Wells Fargo and JPMorgan Chase dominate fracking financing, with Barclays, MUFG, and Mizuho as the biggest funders outside of North America. Fracking is another example of a sector that has devastating local environmental impacts as well as producing new hydrocarbons at a time when we need to be transitioning away from them with considerable urgency.

Liquefied natural gas (LNG): This sector has boomed as policy-makers have bought the fossil industry spiel that gas will serve as a 'transition fuel' rather than the reality that it will delay the transition away from the fossil era. The sector's biggest bankers over the last half decade have been Morgan Stanley, Citi, and JPMorgan Chase.

[1] Banking on Climate Chaos: Fossil Fuel Investment Report 2021. https://www.ran.org/wp-content/uploads/2021/03/Banking-on-Climate-Chaos-2021.pdf.

Coal mining: A sector that is coming under considerable pressure in Europe and the US but is still expanding in Asia, financed by Chinese banks Industrial Bank, China Construction Bank, and Bank of China. On the positive side, the report credits BNP Paribas, BPCE/Natixis, Crédit Mutuel, and UniCredit for limiting investments in coal-mining.

Coal-based electricity generation: This sector is also coming under intense pressure in Europe and the US but new ventures are still funded by Chinese and Japanese banks including Bank of China, ICBC, and China CITIC Bank.

The report closes with some clear demands for banks to fulfil to be considered responsible actors in the climate emergency:

- Reduce the emissions they finance to zero;
- Intermediate commitment to cut;
- With a clear cut significant annual reductions in the emissions they finance from 2021 (for example, NatWest and Lloyds have committed to cutting climate impact in half by 2030);
- Refuse to give credibility to credit and offset schemes that violate human rights, particularly the rights of Indigenous Peoples, are a scam to enable continued production of CO_2 emissions by fossil companies, or depend on technologies that are not yet commercially viable;
- Commitment to dropping clients that don't align with a 1.5°C trajectory;
- Measure and disclose financed emissions.

Beyond the climate crisis, banks are also responsible for financing companies that are causing wider environmental crises that threaten the future of human civilization. In a report called *Bank-Rolling Extinction*,[2] BankTrack report on the environmentally destructive lending activities of some of the world's biggest banks, finding that in 2019, 50 of them provided loans worth US$2.6trn to the economic sectors driving biodiversity loss:

over the course of 2019, 50 global banks together provided loans and underwriting worth more than USD 2.6 trillion to the food, forestry, mining, fossil fuels, infrastructure, tourism and transport and logistics

[2] Bankrolling Extinction: The Top 10 Banks Financing Biodiversity Loss. https://www.banktrack.org/article/bankrolling_extinction_the_top_10_banks_financing_biodiversity_loss.

sectors, all of which have been identified as primary drivers of the global extinction crisis. It also finds that none of the banks have chosen to put comprehensive policies or sufficient systems in place to monitor or measure the impact of their loans on biodiversity.

Unsurprisingly, the roll-call of shame is similar to those found to be bank-rolling the climate crisis: Bank of America, Citigroup, JP Morgan Chase, Mizuho Financial, Wells Fargo, BNP Paribas, Mitsubishi UFJ Financial, HSBC, SMBC Group, and Barclays. And the changes in policy that the banks need to undertake are also similar:

- Banks should disclose the impacts their lending has on natural systems
- Governments should use their role in banking regulation to hold banks liable for the damage caused by their lending.
- People everywhere should have a say in how their money is invested, something that is facilitated by greater transparency via the reporting changes discussed in the previous chapter.

Banks are not like other businesses. They operate under banking licences issues by governments and governments can therefore hold them to higher standards. As discussed later, we need to see banking regulators tighten the screw on commercial banks by excluding fossil assets from counting as collateral against loans and central banks—who operate as the lender of last resort that commercial banks depends on—and refusing to issue licences to banks that continue to lend to fossil fuel corporations.

3 Bank Stability and Sustainability

One of the areas of banking that is opaque to most citizens is the relationship between a bank's capacity to lend and the assets it holds to make good on repaying its depositors. This matters because a proportion of loans will never be repaid (they will 'go bad') but this value must not exceed the money that banks hold to repay depositors. Although we don't generally think about this, everybody who has watched 'It's a Wonderful Life' has a sense of what it means when a bank makes too many promises it can't fulfil. The nature of banking is inherently about walking this tightrope since more loans increase profits but taking too many risks on bad loans puts the bank in jeopardy.

When you study banking in an economics course this balance is taught in terms of the capital ratio which—for the purposes of teaching—is said to be 10%. In reality, that is only done to keep the maths simple. We have all lived through a recent financial crisis, which is the equivalent of all the Bailey Building and Loan banks in the country going bust all at once. That was caused because, over time, the complex financial products that were created overweighted the risk side of the equation compared to the holdings of reliable assets on banks' balance sheets. What regulators insist on today is that banks hold a sufficient volume of risk-weighted assets to maintain both solvency (a bank's capacity to meet its long-term financial commitments) and liquidity (a bank's ability to meet its short-term obligations). What those rates are is determined by an international body called the Basel Committee operating under the auspices of the Bank of International Settlements. For those with a fascination for greater detail or a serious case of insomnia, there is much, much more detail on the BIS website.

We don't need to go into too much detail about how they define which sorts of assets they consider useful collateral and how they assess the risk of various types of assets. Suffice to say that, in a world where the climate and ecological crises are threatening the future returns on most investments and many long-term investments are becoming stranded, these factors need to be taken into account when assessing the riskiness of assets. We made a proposal that this should be made a regulatory requirement in a European Parliament report on Sustainable Finance for which I was rapporteur when we said sustainability risks should be reflected 'in capital requirements and in the prudential consideration of banks' and we asked the Commission 'to adopt a regulatory strategy and a roadmap aimed *inter alia* at measuring sustainability risks within the prudential framework and to promote the inclusion of sustainability risks in the Basel IV framework to ensure sufficient capital reserves'.[3]

[3] Scott Cato report on Sustainable Finance, European Parliament Committee on Economic and Monetary Affairs (2018/2007(INI)), voted in May 2018. https://www.europarl.europa.eu/doceo/document/A-8-2018-0164_EN.html.

In a report focused on greening the UK financial sector,[4] the authors make a similar proposal: 'Conventional risk weights neglect climate issues. We thus propose that risk weights be adjusted to take into account the greenness/dirtiness of the assets that banks hold, based on the UK green/dirty Taxonomy'. (We discussed the Taxonomy in detail in the previous chapter). In other words, bank regulators need to assess assets that relate to economic activity that will play a part in our sustainable future differently from how they assess assets that are becoming or are already stranded.

The authors suggest two ways that this could be undertaken:

> The first form is what is commonly known as the 'dirty penalising factor', which implies an increase in the risk weights of carbon-intensive assets. This intervention would make carbon-intensive lending more expensive relative to low-carbon activities since banks would need to hold more capital against environmentally damaging loans. In this respect, bank lending for carbon intensive activities would be directly dis-incentivised, whilst implicitly encouraging bank lending for low-carbon activities. The second option, a 'green supporting factor', entails a reduction in the risk weight assigned to green assets. Such a reduction would encourage banks to provide more environmentally friendly loans to the economy since banks would have to hold less capital against these loans. It could also lead banks to reduce interest rates on green loans relative to interest rates on conventional loans.

To slightly simplify, when regulators assess the value of a bank's collateral assets they will value those that are not environmentally damaging more highly than those that are.

As the climate and ecological emergencies accelerate it is important that banks regularly test the value of their assets. This is why in my report we called for the introduction of 'carbon stress tests' as had already been proposed by the European Systemic Risk Board (ESRB) in 2016. We went further in suggesting the need for 'climate-resilient prudential policies, such as specific capital adjustment based on the carbon intensity of individual exposures assessed to be excessively applied to the overall investment in assets deemed highly vulnerable to an abrupt transition to

[4] Yannis Dafermos, Daniela Gabor, Maria Nikolaidi and Frank van Lerven (2021), 'Greening the UK Financial System—A Fit for Purpose Approach', *SUERF Policy Note No* 226:9. https://www.suerf.org/docx/f_55c6017b10a9755ef3681b09ccb01e94_21233_suerf.pdf.

the low-carbon economy'. As we have already discussed, climate is only the first crisis that is destroying asset values: over time other unsustainable production practices—including farming—will create financial risks and so stress testing will need to introduce other environmental and social criteria.

My apologies if some of this discussion has become rather technical. In simple summary, it is important to recall that banks create money by making loans and that they can only do this because governments (via their central bank) issue them a banking licence allowing them to do so. The banking licence is dependent on a number of conditions, including conditions about the volume and quality of assets they need to hold against which they make those loans. Now that we understand that the climate and ecological emergencies are destroying the value of some assets (e.g. coal-mines and riverside properties) while increasing the value of others (e.g. wind-farms and organic farms) it becomes clear that banking regulatory authorities need to include consideration of the sustainability risks of bank assets when they decide whether banks are adequately capitalized. They are beginning to do that and it will transform the value of assets in the wider economy, as many cease to be viable as bank collateral.

4 Central Banks as a Force for Good?

Greens have long challenged the political consensus that central banks should be politically neutral—refusing to accept that the extraordinary power to create money should be exercised in secret, for the benefit of a tiny minority, and often at the expense of the planet. In a time of such existential crisis, why would you allow one of the most powerful forces at your disposal—the power to create and allocate finance—to remain on the political sidelines. It is a rhetorical question, of course, because the answer is about the power of vested interests. But it is a question that is being increasingly asked and with some indication that the repetition is bringing movement.

The political case is made clearly in a report from the New Economics Foundation: 'In modern economies, the banking system creates between 85 and 97% of the money supply. Whilst governments and non-bank financial intermediaries – like pension funds – tax and spend existing money, banks create new money and purchasing power via the act of lending. At the aggregate level, their lending decisions have the power to shape the long-term trajectory of the economy'.[5] And in a climate and ecological emergency, they should be bending that trajectory towards sustainability. In this section we will investigate suggestions for how they might do this.

A report from the LSE's Grantham Centre makes suggestions in a number of areas. Firstly, central banks need to mainstream the climate crisis into their strategic thinking, producing a climate-neutrality roadmap and forward planning scenarios consistent with a 1.5°C warming limit. They also propose the need for climate change to be taken into account in prudential regulation and in banks' monetary policy, something we will discuss in more detail shortly, together with the fact that the assets they hold themselves need to be carbon–neutral. Finally, the authors note the importance of ensuring that we undergo a just transition to a sustainable future—with banking lending being used to minimize risks that people or regions get left behind—and that central banks across the world cooperate on this vital agenda.[6]

David Barmes, lead economist at thinktank Positive Money, has produced a scorecard assessing the relative sustainability performance of different central banks around the world (reproduced in Fig. 1).[7] This report makes an assessment on four different categories: Research and Advocacy, Monetary Policy, Financial Policy, and Leading by Example. Most central banks score well on the first category but much less well on the others, suggesting that the rhetoric is currently outstripping performance. Intriguingly, China heads the table, closely followed by Brazil and France and then the UK and the ECB. The US barely scores at all in terms of actions taken. The authors conclude that:

[5] Frank van Lerven and Josh Ryan-Collins (2017), Central Banks, Climate Change, and the Transition to a Low-Carbon Economy, (London: nef); McLeay, M., Radia, A. and Thomas, R. (2014). 'Money Creation in the Modern Economy', *Bank of England Quarterly Bulletin 2014 Q1*. London: Bank of England.

[6] Simon Dikau, Nick Robins, Ulrich Volz (2021), *Climate-Neutral Central Banking: How the European System of Central Banks Can Support the Transition to Net-zero* (London: Grantham Institute). https://www.lse.ac.uk/granthaminstitute/publication/climate-neutral-central-banking-how-the-european-system-of-central-banks-can-support-the-transition-to-net-zero/.

[7] David Barmes and Zack Livingstone (2021), *The Green Central Banking Scorecard: How Green are the G20 Central Banks and Financial Supervisors?* (London: Positive Money). https://positivemoney.org/publications/green-central-banking-scorecard/.

> Brazil's deforestation is rampant, while China is renowned for its high levels of air, water and soil pollution. Nevertheless, these countries' institutions are ranked above those in multiple other countries that have less severe environmental crises within their borders. Why are these countries' institutions performing better than multiple others that appear to have less severe environmental crises within their borders? China and Brazil are both rich in natural resources, and significant environmental harm has been generated in the exploitation of these resources, driven in large part by overconsumption within the high-income economies of countries in the Global North... This makes environmental impacts and risks more immediately visible and relevant for their central bankers and supervisors, and may result in a greater impetus to green their policymaking processes.

They also note that this is a new agenda for central banks, acknowledge progress, and anticipate rapid progress in the next few years.

Partly as a result of campaigning like this, central banks are sitting up and taking notice and accelerating their focus on their role in the climate and ecological emergencies. In 2017, representatives from some of the central banks leading on this agenda—China, France, Germany, Mexico, Singapore, Sweden, the UK, and the Netherlands—founded the Central Banks and Supervisors Network for Greening the Financial System (NGFS), at a meeting held under the auspices of the Dutch National Bank, Banque de France, and Bank of England. The Network responds directly to the Paris Agreement of 2015 and aims 'to manage risks and to mobilize capital for green and low-carbon investments in the broader context of environmentally sustainable development'.[8]

There were some encouraging signs in this year's UK Budget that the Bank of England may finally be moving in this direction and might use *my* national currency to support rather than undermine the sustainability transition. Following a long campaign led by Positive Money and others, the Chancellor changed the Bank's mandate so that it now has to align its lending with the government's net-zero target so that it will 'reflect the importance of environmental sustainability and the transition to net zero'.[9]

[8] https://www.ngfs.net/en/about-us/governance/origin-and-purpose.

[9] https://assets.publishing.service.gov.uk/government/uploads/system/uploads/attachment_data/file/965782/2021_MPC_remit_FINAL_1_March_.pdf.

Rank	Country	Research and Advocacy (out of 10)	Monetary Policy (out of 50)	Financial Policy (out of 50)	Leading by Example (out of 20)	Aggregate Score (out of 130)	Grade (A+ to F)
1	China	10	16	24	0	50	C
2	Brazil	10	16	18	1	45	C-
3	France	10	3	22	8	43	C-
4	United Kingdom	10	4	19	5	38	D+
5	European Union	10	2	15	6	33	D+
6	Italy	10	2	15	4	31	D+
7	Germany	10	1	15	3	29	D
8	Indonesia	10	1	8	2	21	D
9	Japan	10	5	4	0	19	D-
10	Australia	10	0	4	1	15	D-
11=	Canada	10	0	2	1	13	D-
11=	Mexico	10	1	1	1	13	D-
13=	South Korea	10	0	1	0	11	D-
13=	United States	10	0	1	0	11	D-
15	India	3	5	1	0	9	F
16	Russia	5	0	1	2	8	F
17	South Africa	7	0	0	0	7	F
18	Turkey	1	0	3	0	4	F
19=	Argentina	0	0	0	0	0	F
19=	Saudi Arabia	0	0	0	0	0	F

Fig. 1 Central bank sustainability performance scorecard (*Note* G20 countries ranked by green monetary and financial policies. *Source* Thanks to David Barmes at Positive Money for permission to reproduce this graphic free of charge)

Experts and campaigners hope this mandate reform will translate to key changes to Bank of England policies, such as:

- Excluding bonds issues by fossil fuel companies from its purchasing scheme (QE—see more below);
- Ensuring that the carbon intensity of the portfolio as a whole is consistent with the Paris Agreement;
- Decarbonizing the assets that the Bank holds as collateral for its own lending;
- Using its power to create money to invest in sustainable jobs via the National Infrastructure Bank;
- Using the Term Funding Scheme to incentivize lending to green SMEs.

These policy shifts could be transformational in ensuring that the power of the UK's national currency is used to accelerate the sustainability transition.

But central banks also need to use their money-creation powers in a pro-environment way to play their full part in achieving sustainable finance. The whole idea of 'quantitative easing', using the power of a central bank to inject money into the economy, was dismissed as marginal and unworkable until the 2008 financial crises made it inevitable and mainstream. But this extraordinary power to directly stimulate some sectors rather than others has not been used to accelerate the sustainability transition. In fact, quite the reverse:

> Central banks have expanded their monetary policy interventions significantly in the face of economic stagnation following the financial crisis of 2008. New money has been created – 'printed' in the pre-digital terminology – and pumped into the economy to stimulate the purchase of assets and thus, indirectly, wider spending. The world's four major central banks have expanded their balance sheets on average from 10% of GDP in 2008 to 45% today. However, central banks have generally not aligned their policy objectives with the threats of climate change. Indeed, some central bank policy is even having unintended negative implications for the environment.[10]

[10] Frank van Lerven and Josh Ryan-Collins (2017), Central Banks, Climate Change, and the Transition to a Low-Carbon Economy, (London: nef), p. 8.

For this reason, in 2016 I wrote with other Green MEPs to Mario Draghi, then President of the European Central Bank that he was responsible for regulating, requesting him to stop the purchase of fossil fuel and other unsustainable assets as part of the bank's Corporate Sector Purchase Programme and asking him to ensure 'full alignment with EU environmental global commitments of the ECB's current or, eventually in future, private sector purchase programmes'. I continued to lobby him to shift the purchases towards socially and environmentally beneficial sectors. The full potential of Green QE was outlined as long ago as 2015 in a report I commissioned from Victor Anderson.[11]

I'm pleased to say that these efforts have borne fruit with the ECB's strategy review, published in July 2021, making it clear that they will align their role as European central bank more closely with the needs of European citizens and with the European Green Deal. In future, both its modelling of scenarios and its assessment of risks will take account of climate change. It will take account of sustainability impacts when making its assets purchases and has also committed to undertaking stress testing of the climate impact of its asset holdings.[12]

As well as having the power to issue currency, central banks are also responsible for the regulation of the private and public banks that issue currency within their jurisdictions. In other words, they set the rules within which banks hold a licence and can use that licence to—almost literally—print money. They need to regulate in a way that minimizes risks and that includes risks from the carbon bubble. This is what is unattractively known as 'macro-prudential regulation'.

So what might it mean in practice? Frank van Lerven and Josh Ryan-Collins suggest that they might increase 'the commercial banking sector's capital requirements, e.g. by forcing banks to hold a higher portion of capital against certain types of loans they make... Similarly, macroprudential policy may involve implementing quantitative limits on certain type of banks loans'. They make the case in terms of unsustainable lending in an

[11] Greens-EFA Group (2015), Green Money: Reclaiming Quantitative Easing Money Creation for the Common Good, Written by Victor Anderson; Commissioned by Molly Scott Cato MEP; Funded and published by the Green/EFA group in the European Parliament.

[12] Website of MEP Sven Giegold. https://sven-giegold.de/en/ecb-monetary-policy-greener/.

overheated housing market, but it could be made just as well in the case of environmentally unsustainable lending.

Further down the track we have a policy known as 'credit guidance', meaning that the power to create money, and therefore direct economic activity, should be used to ensure we accelerate the sustainability transition. Perhaps the Bank of England could set a declining proportion of bank lending for fossil sectors and set a minimum proportion for sustainable sectors, for example. While the changes we are seeing at central banks have not gone as far as such a policy of credit guidance, a significant number of such banks are moving towards a position where they should prioritize using their money-creation powers to actively support the sustainability transition. Van Lerven and Ryan-Collins note that this principle of directing credit flows towards sector crucial to the public good was used widely in Western countries in the post-war period and supported the expansion of East Asian economies—including that of China—from the 1970s onwards. They suggest a number of mechanisms:

- Limits on 'brown' lending or quotas for green lending:
- Green refinancing: allowing commercial banks to borrow from the central bank ('refinance') at lower rates to ease financing constraints in green sectors and to encourage banks to lend for green purposes.
- Green reserve requirements: requiring lower reserve holdings for banks lending more for sustainable activities.

Our rapidly heating climate is a national and global emergency and it is absurd to suggest that we cannot afford to solve it. Conversely, we should be using the power of finance, through our central banks, to send that finance towards the sectors we need as we transition to a sustainable future.

5 The Leadership Role of Public Banks

We have considered the role of private banks and central banks now we move on to the third important category of banks that have a crucial role to play in sustainable finance: public banks. Public banks have the same powers to create money through lending as private banks but they are owned by a public body such as a national or regional government. This means that they can invest the money they lend for public benefit and that the profits they make on lending are also reinvested according to social priorities, rather than being paid as dividends to shareholders. Public banks can also finance innovative markets that are not well enough establish to attract the confidence of private financiers.

Having a requirement to lend in the public interest means that public banks can mirror public policy and priorities of state, regional, or supranational governments, making the finance available to enable the projects and infrastructure they require. The European Investment Bank is an example of such a bank—owned by the EU's member states—that is now able to finance the European Green Deal. The EIB has led in the area of green finance and issued the world's first green bond in 2007. The EIB undertook its role as a public bank in supporting the offshore wind sector before it became a stable and mature market. This has included providing over €3 billion for developing and expanding the UK's offshore wind industry.[13]

The EIB has faced some criticism in recent years over its investment decisions, some of which have locked fossil fuels into the European economy, notoriously so in the case of the €1.5bn loan for the Trans-Adriatic Pipeline (TAP) that will cross Northern Greece, Albania, and the Adriatic Sea before coming ashore in Southern Italy to connect to the Italian natural gas network. However, in November 2019 it rebranded itself as Europe's Climate Bank with President Werner Hoyer pledging to end all financing of fossil fuels and use their resources instead to power the European Green Deal.[14]

The most impressive public bank I know about is Germany's KfW (Kreditanstalt für Wiederaufbau or 'Reconstruction Credit Institute'), which calls itself a public development bank, akin to banks operating in emerging economies, on the understanding that certain sectors are always emerging even in the high-income economies. It was founded to provide finance to rebuild the German economy after World War II, then went on to fund the poorer parts of Germany after Reunification, and is now providing finance for the sustainability transition. One arm of KfW (KfW Development Bank) operates as a global development bank, providing

[13] House of Lords European Union Committee (2019), Brexit: the European Investment Bank, 25th Report of Session 2017–19, HL Paper 269. https://publications.parliament.uk/pa/ld201719/ldselect/ldeucom/269/26905.htm.

[14] https://www.eib.org/en/press/all/2019-313-eu-bank-launches-ambitious-new-climate-strategy-and-energy-lending-policy.

finance in some of the world's poorest countries on behalf of the German Federal Government. These investments follow the criteria of the UN's 2030 Agenda for Sustainable Development, the 17 Sustainable Development Goals (SDGs) and the Paris Agreement on climate change. In 2019, KfW Development Bank committed €8.8bn across the world in this way.[15]

Far from the image of shareholders and pinstripe suits, this 'bank' is a powerhouse of sustainability across the world. KfW has been instrumental in financing Germany's energy transition and is empowered to invest especially in economically depressed areas. As a public bank, interest payments can be reinvested for public benefit rather than extracted by wealthy shareholders. In 2019, it supplied funds totalling €77.3bn, 38% of which was spent on measures to protect the climate or the environment. KfW was one of the first five institutions to be accredited by the Green Climate Fund (GCF) in 2015 for the implementation of grant financing.[16] KfW's first project with the GCF was launched in Bangladesh providing a grant of $40 million to fund a project to improve protection for infrastructure from the consequences of climate change.

Germany's banking system—its regional and local cooperation banks, as well as its public banks of which KfW is the largest—are a key reason for the success of the German economy, but one that is rarely discussed. So KfW lends money to young entrepreneurs at favourable rates especially in the areas of renewables and energy efficiency. It also makes loans available for home retrofits and age-appropriate conversion and offers favourable mortgage rates for energy-efficient properties or for homeowners who carry out conversions or renovations to reduce energy use.

I can only apologize for this slightly breathless account of the triumph that is KfW. In my last week as an MEP I gave myself the treat of an in-person visit and have to confess to being something of a superfan! It is a source of regret that the UK has never had a public development bank, forcing public authorities and social businesses to borrow at commercial rates. Rather than the UK expanding public banking facilities they are currently shrinking. The Green Investment Bank, established in 2012, was really a revolving loan fund considered a drain on the public finances

[15] KfW at a Glance: Facts and Figures. https://www.kfw.de/PDF/Download-Center/Konzernthemen/KfW-at-a-glance/KfW-an-overview.pdf.

[16] https://www.kfw-entwicklungsbank.de/International-financing/KfW-Development-Bank/News/News-Details_460160.html.

by Chancellor George Osborne. It was privatized and sold to Australian commercial bank Macquarie in 2017. In the spring of 2021, the Chancellor announced the creation of a UK infrastructure bank with initial capitalization of £12 billion and many are watching this development with interest. It goes without saying that, because of the UK leaving the EU, government, and public authorities no longer have access to EIB funding.

Pro-sustainability thinktank E3G has produced a comparison of the performance of various public development banks, which they assess to be responsible for $2.3trn on investment around the world, some 10% of total global finance. It's a valuable tool to assess how banks are doing on a range of different aspects of sustainability according to the indicators as shown in Fig. 2.

From their work we learn that the Asian Development Bank gets tops marks on nature-based solutions having worked in depth on the Greater Mekong region and the field water of Vietnam as well as making a peatlands assessment of Mongolia. Or that the World Bank is in the red in terms of its performance on energy-efficiency strategy, standards and investment. This poor rating is because it has does not have energy-efficiency standards except for transport. And we learn that the Inter-American Development Bank scores poorly on energy access and fuel poverty because, although it has set a clear goal, it is not making good progress or monitoring adequately: the bank has committed to the Sustainable Energy for All initiative in the Americas but it is not clear what progress is being made to connect the last 30 million people without access to electricity.

Obviously, I am only able to give a sense of the depth of information available in the tracker matrix but it gives a flavour of the expectations on such banks and monitoring they are now subject to, and which can only increase.

6 World's Bank Missing in Action

We've considered the role of national and supranational banks, but what about the world's bank? What role is it, or should it be, playing in this greatest crisis ever to threaten the world, in human terms at least? Given that 190 of the world's 196 countries are members, shouldn't we be expecting more action?

Our international financial institutions date back to the post-war settlement and have their origins in the 1944 Bretton Woods conference.

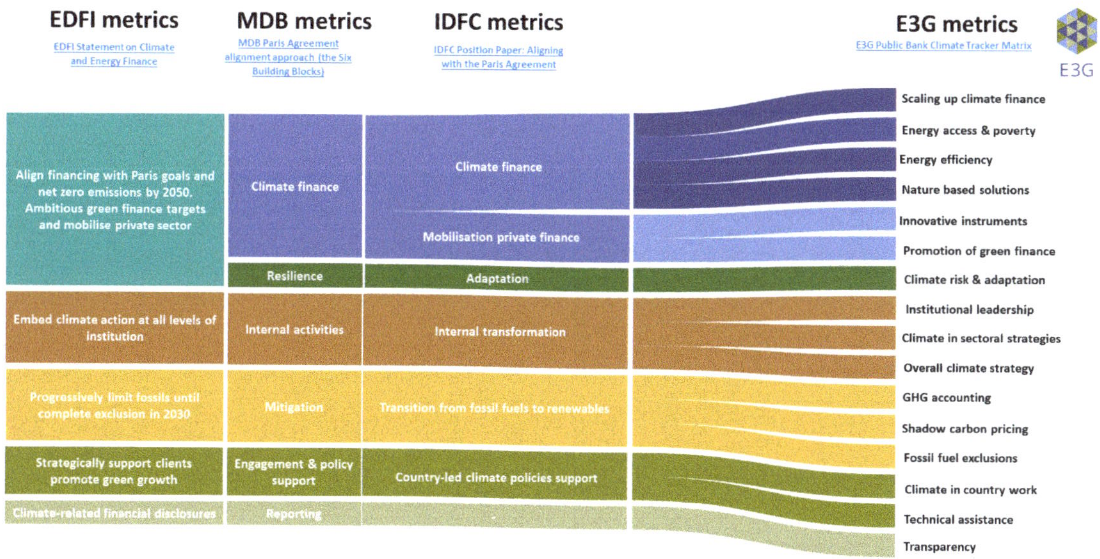

Fig. 2 E3G's 15 metrics of Paris agreement alignment at public and development banks (*Note* This is part of the E3G Public Bank Climate Tracker Matrix: https://www.e3g.org/matrix/. *Source* Thanks to James Hawkins and Sonia Dunlop at E3G)

As in much discussion around finance and economics, their names are misleading, since the International Monetary Fund is really a bank and the World Bank operates more like an investment fund. I say that because the role of World Bank has been to make grants and loans to countries for capital and infrastructure projects. The IMF has the power to issue liquidity in the form of SDRs (Special Drawing Rights) so has the monetary creation characteristic of a bank. Their role has always been primarily to ensure the stability of the international financial system, but they have also been involved in the 'development' of countries of the Global South—some would say positively, others not. So my question, that I go into in more detail in Chapter 3, is why isn't the IMF using that power to create money to solve the climate crisis? And generally, why are the global financial institutions showing so little leadership on the climate and ecological crises?

Shortly after taking over as managing director of the IMF in October 2020, Kristalina Georgieva made some ambitious pledges in an interview with *Time* magazine, with warm words about a green recovery from Covid. However, like so many top bankers she seemed to focus on using regulatory powers around risks to financial stability and carbon stress testing rather than the monetary powers of her organization.[17] Speaking at a conference in Vienna in July 2021 she began by mentioning the IMF's Climate Change Indicators Dashboard before hinting that the bank might begin to use its currency-issuing powers, 'the next step is to consider whether and how IMF financing can help implement policy advice – including climate mitigation, adaptation, and transition policies. In the context of the forthcoming allocation of $650 billion of Special Drawing Rights, we are exploring the creation of a Resilience and Sustainability Trust'.[18] (see more on how this might help a solution to the climate crisis in Chapter 3.)

Here I will just explain a bit more about SDRs. As explained earlier, central banks are responsible for issuing reserve currencies and they hold

[17] IMF Chief Kristalina Georgieva Says We Have a 'Once in a Century Opportunity' to Fund a Sustainable Future, interview by Justin Worland, *Time*, October 21, 2020: https://time.com/collection/great-reset/5901323/imf-kristalina-georgieva-sustainable-economy/.

[18] Remarks by IMF 'Managing Director on Global Policies and Climate Change', *International Conference on Climate*, Venice, 11 July 2021. https://www.imf.org/en/News/Articles/2021/07/11/sp071121-md-on-global-policies-and-climate-change.

reserve assets (originally gold but now a 'basket' of other countries' reliable reserve currencies and, overwhelmingly, dollars). At Bretton Woods, it was agreed that the US dollar would operate as the global reserve asset—as it still does—but in 1969 the IMF agreed to issue its own international reserve assets, the Special Drawing Right or SDR. SDRs are not currency themselves but rather a potential claim on the currencies of IMF members, so they can be exchanged freely for US dollars, for example. Their value is based on the value of a basket of five global currencies: the US dollar, the euro, the Chinese renminbi, the Japanese yen, and the British pound sterling.

The IMF can issue these to stabilize the global financial system if liquidity dries up, as it did after the 2008 financial crisis. In August 2021, the largest ever allocation of these SDRs was made to help cope with recovery from the Covid-19 pandemic. Given the power of this it really is surprising that it was not headline news around the world. This probably sounds too good to be true and you may be asking what is the catch. There most certainly is one and that is that SDRs are not allocated on a per capita basis to the world's citizens but according to the shares that countries hold in the IMF. This means that 16.5% of the extraordinary largesse went to the US, 6% to China, and 4% to the UK, but only 0.5% to Nigeria, 0.38% to Chile, and 0.09% to Afghanistan. It seems that the principles governing the IMF are not those of global equity but of the parable of the talents: 'For unto every one that hath shall be given, and he shall have abundance: but from him that hath not shall be taken away even that which he hath'.

Elsewhere I have argued the importance of restructuring the global financial institutions so that the world can have a bank that operates in all our interests and to protect the planet we share.[19] Green economist Richard Douthwaite proposed embedding climate responsibility into the global reserve currency through the creation of an EBCU, an energy-backed currency unit.[20] The global central bank would be responsible for exchange rate control and would act as a lender of last resort to countries experiencing reserve crises. It would be backed by deposits from member countries and voting rights would be based on population shares rather

[19] Cato, M. S. (2009a), 'A New Financial Architecture based on a Global Carbon Standard', *Ecopolitics*, 3:61–78; reprinted in Barry, J. and Leonard, L. (eds.), *Advances in Ecopolitics*, p. iii.

[20] Douthwaite, R. (1999), The Ecology of Money (Totnes: Green Books), p. 57.

than deposits. It would also take responsibility for the issuing of EBCUs to countries, which can be used for trade or to buy carbon permits. The concept is that the value of the global climate—which is, after all, invaluable—would be used to support the issue of the global currency and that issue could be used to ensure climate action by governments. Whether this proposal or something like it is adopted or not, we have a right to look to the world's bank to fund the survival of the human species and their actions have been woefully inadequate thus far.

So we've reached the end of the whistle-stop tour around the world of banking. I hope it was more interesting than you expected. So many of the secrets about why there is poverty and environmental crisis lie in understanding how money and banking operate in our global economy that I hope you will follow up some of the links and continue learning more about this area. I have to say that, not unlike trainspotting, you may find yourself getting hooked.

Epilogue

This was always intended to be a short book but I hope I have covered the basics of the key issues in the sustainable finance agenda. I have tried to do so in an approachable way and to provide links to further and deeper explorations in every chapter. These make it possible for a reader to follow up on any particular area of interest and to seek out further sources to ensure you have a deeper and more technical understanding.

I have already shared the view of commentators that, rather than sharing objectives or even a view of the world, those who have worked to advance the cause of sustainable finance are a coalition of disparate elements. But if what we share is a desire to preserve our beautiful planet and the human species' ability to exist on it then this is surely a noble endeavour even if their objectives and philosophies differ greatly. In some senses my greatest joy in being involved in this agenda is that I have been able to work with those whose view of the world and a good life is so distant from my own but in spite of these we have found ways to advance the sustainable finance agenda across our gulfs of understanding and ideology.

The journey of sustainable finance is only just beginning. I have tried to explain in this book areas where we have made progress and also areas where there is much further work to be done. In the area of climate, significant advances have been made in terms of pricing the risk associated with stranded assets and seeking ways to favour investments compatible

© The Author(s), under exclusive license to Springer Nature Switzerland AG 2022

M. Scott Cato, *Sustainable Finance*,
https://doi.org/10.1007/978-3-030-91578-0

with the Paris Agreement. I believe this can create a prototype for regulating the sustainability of investments in other areas but there is still space for creative thinking and for those who understand the world of finance to prioritize work using that knowledge to preserve our life-support systems rather than to maximize profits. I would invite any experts in finance non-financial accounting or banking who have read this book as an introduction to join us on this adventure because nothing can be more important right now.

Index

© The Author(s), under exclusive license to Springer Nature Switzerland AG 2022
M. Scott Cato, *Sustainable Finance*,
https://doi.org/10.1007/978-3-030-91578-0

Lightning Source UK Ltd.
Milton Keynes UK
UKHW021820270122
397804UK00001B/52

9 783030 915773